CHST Study Guide

CHST Exam Prep & Practice Test Questions for the Construction Health & Safety Technician Exam
[3rd Edition]

Test Prep Books

Written and edited by Test Prep Books.

ISBN 13: 9781628457018
ISBN 10: 1628457015

Table of Contents

Quick Overview

As you draw closer to taking your exam, effective preparation becomes more and more important. Thankfully, you have this study guide to help you get ready. Use this guide to help keep your studying on track and refer to it often.

This study guide contains several key sections that will help you be successful on your exam. The guide contains tips for what you should do the night before and the day of the test. Also included are test-taking tips. Knowing the right information is not always enough. Many well-prepared test takers struggle with exams. These tips will help equip you to accurately read, assess, and answer test questions.

A large part of the guide is devoted to showing you what content to expect on the exam and to helping you better understand that content. In this guide are practice test questions so that you can see how well you have grasped the content. Then, answer explanations are provided so that you can understand why you missed certain questions.

Don't try to cram the night before you take your exam. This is not a wise strategy for a few reasons. First, your retention of the information will be low. Your time would be better used by reviewing information you already know rather than trying to learn a lot of new information. Second, you will likely become stressed as you try to gain a large amount of knowledge in a short amount of time. Third, you will be depriving yourself of sleep. So be sure to go to bed at a reasonable time the night before. Being well-rested helps you focus and remain calm.

Be sure to eat a substantial breakfast the morning of the exam. If you are taking the exam in the afternoon, be sure to have a good lunch as well. Being hungry is distracting and can make it difficult to focus. You have hopefully spent lots of time preparing for the exam. Don't let an empty stomach get in the way of success!

When travelling to the testing center, leave earlier than needed. That way, you have a buffer in case you experience any delays. This will help you remain calm and will keep you from missing your appointment time at the testing center.

Be sure to pace yourself during the exam. Don't try to rush through the exam. There is no need to risk performing poorly on the exam just so you can leave the testing center early. Allow yourself to use all of the allotted time if needed.

Remain positive while taking the exam even if you feel like you are performing poorly. Thinking about the content you should have mastered will not help you perform better on the exam.

Once the exam is complete, take some time to relax. Even if you feel that you need to take the exam again, you will be well served by some down time before you begin studying again. It's often easier to convince yourself to study if you know that it will come with a reward!

Test-Taking Strategies

1. Predicting the Answer

When you feel confident in your preparation for a multiple-choice test, try predicting the answer before reading the answer choices. This is especially useful on questions that test objective factual knowledge. By predicting the answer before reading the available choices, you eliminate the possibility that you will be distracted or led astray by an incorrect answer choice. You will feel more confident in your selection if you read the question, predict the answer, and then find your prediction among the answer choices. After using this strategy, be sure to still read all of the answer choices carefully and completely. If you feel unprepared, you should not attempt to predict the answers. This would be a waste of time and an opportunity for your mind to wander in the wrong direction.

2. Reading the Whole Question

Too often, test takers scan a multiple-choice question, recognize a few familiar words, and immediately jump to the answer choices. Test authors are aware of this common impatience, and they will sometimes prey upon it. For instance, a test author might subtly turn the question into a negative, or he or she might redirect the focus of the question right at the end. The only way to avoid falling into these traps is to read the entirety of the question carefully before reading the answer choices.

3. Looking for Wrong Answers

Long and complicated multiple-choice questions can be intimidating. One way to simplify a difficult multiple-choice question is to eliminate all of the answer choices that are clearly wrong. In most sets of answers, there will be at least one selection that can be dismissed right away. If the test is administered on paper, the test taker could draw a line through it to indicate that it may be ignored; otherwise, the test taker will have to perform this operation mentally or on scratch paper. In either case, once the obviously incorrect answers have been eliminated, the remaining choices may be considered. Sometimes identifying the clearly wrong answers will give the test taker some information about the correct answer. For instance, if one of the remaining answer choices is a direct opposite of one of the eliminated answer choices, it may well be the correct answer. The opposite of obviously wrong is obviously right! Of course, this is not always the case. Some answers are obviously incorrect simply because they are irrelevant to the question being asked. Still, identifying and eliminating some incorrect answer choices is a good way to simplify a multiple-choice question.

4. Don't Overanalyze

Anxious test takers often overanalyze questions. When you are nervous, your brain will often run wild, causing you to make associations and discover clues that don't actually exist. If you feel that this may be a problem for you, do whatever you can to slow down during the test. Try taking a deep breath or counting to ten. As you read and consider the question, restrict yourself to the particular words used by the author. Avoid thought tangents about what the author *really* meant, or what he or she was *trying* to say. The only things that matter on a multiple-choice test are the words that are actually in the question. You must avoid reading too much into a multiple-choice question, or supposing that the writer meant something other than what he or she wrote.

5. No Need for Panic

It is wise to learn as many strategies as possible before taking a multiple-choice test, but it is likely that you will come across a few questions for which you simply don't know the answer. In this situation, avoid panicking. Because most multiple-choice tests include dozens of questions, the relative value of a single wrong answer is small. As much as possible, you should compartmentalize each question on a multiple-choice test. In other words, you should not allow your feelings about one question to affect your success on the others. When you find a question that you either don't understand or don't know how to answer, just take a deep breath and do your best. Read the entire question slowly and carefully. Try rephrasing the question a couple of different ways. Then, read all of the answer choices carefully. After eliminating obviously wrong answers, make a selection and move on to the next question.

6. Confusing Answer Choices

When working on a difficult multiple-choice question, there may be a tendency to focus on the answer choices that are the easiest to understand. Many people, whether consciously or not, gravitate to the answer choices that require the least concentration, knowledge, and memory. This is a mistake. When you come across an answer choice that is confusing, you should give it extra attention. A question might be confusing because you do not know the subject matter to which it refers. If this is the case, don't eliminate the answer before you have affirmatively settled on another. When you come across an answer choice of this type, set it aside as you look at the remaining choices. If you can confidently assert that one of the other choices is correct, you can leave the confusing answer aside. Otherwise, you will need to take a moment to try to better understand the confusing answer choice. Rephrasing is one way to tease out the sense of a confusing answer choice.

7. Your First Instinct

Many people struggle with multiple-choice tests because they overthink the questions. If you have studied sufficiently for the test, you should be prepared to trust your first instinct once you have carefully and completely read the question and all of the answer choices. There is a great deal of research suggesting that the mind can come to the correct conclusion very quickly once it has obtained all of the relevant information. At times, it may seem to you as if your intuition is working faster even than your reasoning mind. This may in fact be true. The knowledge you obtain while studying may be retrieved from your subconscious before you have a chance to work out the associations that support it. Verify your instinct by working out the reasons that it should be trusted.

8. Key Words

Many test takers struggle with multiple-choice questions because they have poor reading comprehension skills. Quickly reading and understanding a multiple-choice question requires a mixture of skill and experience. To help with this, try jotting down a few key words and phrases on a piece of scrap paper. Doing this concentrates the process of reading and forces the mind to weigh the relative importance of the question's parts. In selecting words and phrases to write down, the test taker thinks about the question more deeply and carefully. This is especially true for multiple-choice questions that are preceded by a long prompt.

9. Subtle Negatives

One of the oldest tricks in the multiple-choice test writer's book is to subtly reverse the meaning of a question with a word like *not* or *except*. If you are not paying attention to each word in the question, you can easily be led astray by this trick. For instance, a common question format is, "Which of the following is...?" Obviously, if the question instead is, "Which of the following is not...?," then the answer will be quite different. Even worse, the test makers are aware of the potential for this mistake and will include one answer choice that would be correct if the question were not negated or reversed. A test taker who misses the reversal will find what he or she believes to be a correct answer and will be so confident that he or she will fail to reread the question and discover the original error. The only way to avoid this is to practice a wide variety of multiple-choice questions and to pay close attention to each and every word.

10. Reading Every Answer Choice

It may seem obvious, but you should always read every one of the answer choices! Too many test takers fall into the habit of scanning the question and assuming that they understand the question because they recognize a few key words. From there, they pick the first answer choice that answers the question they believe they have read. Test takers who read all of the answer choices might discover that one of the latter answer choices is actually *more* correct. Moreover, reading all of the answer choices can remind you of facts related to the question that can help you arrive at the correct answer. Sometimes, a misstatement or incorrect detail in one of the latter answer choices will trigger your memory of the subject and will enable you to find the right answer. Failing to read all of the answer choices is like not reading all of the items on a restaurant menu: you might miss out on the perfect choice.

11. Spot the Hedges

One of the keys to success on multiple-choice tests is paying close attention to every word. This is never truer than with words like almost, most, some, and sometimes. These words are called "hedges" because they indicate that a statement is not totally true or not true in every place and time. An absolute statement will contain no hedges, but in many subjects, the answers are not always straightforward or absolute. There are always exceptions to the rules in these subjects. For this reason, you should favor those multiple-choice questions that contain hedging language. The presence of qualifying words indicates that the author is taking special care with his or her words, which is certainly important when composing the right answer. After all, there are many ways to be wrong, but there is only one way to be right! For this reason, it is wise to avoid answers that are absolute when taking a multiple-choice test. An absolute answer is one that says things are either all one way or all another. They often include words like *every*, *always*, *best*, and *never*. If you are taking a multiple-choice test in a subject that doesn't lend itself to absolute answers, be on your guard if you see any of these words.

12. Long Answers

In many subject areas, the answers are not simple. As already mentioned, the right answer often requires hedges. Another common feature of the answers to a complex or subjective question are qualifying clauses, which are groups of words that subtly modify the meaning of the sentence. If the question or answer choice describes a rule to which there are exceptions or the subject matter is complicated, ambiguous, or confusing, the correct answer will require many words in order to be expressed clearly and accurately. In essence, you should not be deterred by answer choices that seem excessively long. Oftentimes, the author of the text will not be able to write the correct answer without offering some qualifications and modifications. Your job is to read the answer choices thoroughly and completely and to select the one that most accurately and precisely answers the question.

13. Restating to Understand

Sometimes, a question on a multiple-choice test is difficult not because of what it asks but because of how it is written. If this is the case, restate the question or answer choice in different words. This process serves a couple of important purposes. First, it forces you to concentrate on the core of the question. In order to rephrase the question accurately, you have to understand it well. Rephrasing the question will concentrate your mind on the key words and ideas. Second, it will present the information to your mind in a fresh way. This process may trigger your memory and render some useful scrap of information picked up while studying.

14. True Statements

Sometimes an answer choice will be true in itself, but it does not answer the question. This is one of the main reasons why it is essential to read the question carefully and completely before proceeding to the answer choices. Too often, test takers skip ahead to the answer choices and look for true statements. Having found one of these, they are content to select it without reference to the question above. Obviously, this provides an easy way for test makers to play tricks. The savvy test taker will always read the entire question before turning to the answer choices. Then, having settled on a correct answer choice, he or she will refer to the original question and ensure that the selected answer is relevant. The mistake of choosing a correct-but-irrelevant answer choice is especially common on questions related to specific pieces of objective knowledge. A prepared test taker will have a wealth of factual knowledge at his or her disposal, and should not be careless in its application.

15. No Patterns

One of the more dangerous ideas that circulates about multiple-choice tests is that the correct answers tend to fall into patterns. These erroneous ideas range from a belief that B and C are the most common right answers, to the idea that an unprepared test-taker should answer "A-B-A-C-A-D-A-B-A." It cannot be emphasized enough that pattern-seeking of this type is exactly the WRONG way to approach a multiple-choice test. To begin with, it is highly unlikely that the test maker will plot the correct answers according to some predetermined pattern. The questions are scrambled and delivered in a random order. Furthermore, even if the test maker was following a pattern in the assignation of correct answers, there is no reason why the test taker would know which pattern he or she was using. Any attempt to discern a pattern in the answer choices is a waste of time and a distraction from the real work of taking the test. A test taker would be much better served by extra preparation before the test than by reliance on a pattern in the answers.

Introduction to the CHST

Function of the Test

The Construction Health and Safety Technician (CHST) examination is part of the CHST certification process administered by the Board of Certified Safety Professionals (BCSP). The certification process is intended to identify and validate the level of competency of individuals whose work involves health and safety activities relating to the prevention of construction-related illnesses and injuries. A CHST examination score that meets the minimum passing level permits such an individual to receive CHST certification and the title of Construction Health and Safety Technician.

CHST exam candidates must have a minimum of three years of construction experience with at least 35% of their primary job responsibilities relating to safety and health. Accordingly, test takers are typically already working in the construction industry in positions relating to safety and health on construction projects or job sites. Individuals who pass the CHST exam and obtain CHST certification can use it to progress in the field to more significant safety and health responsibilities and/or as a step on the way to obtaining a Certified Safety Professional certification.

Test Administration

The CHST examination is offered at Pearson VUE testing centers across the country and is typically available on any and all business days for which a given center is open. Prospective test takers must apply through BCSP. Once a test taker's application is approved, he or she has one year to pay the registration fee, schedule the exam through Pearson VUE, and sit for the exam.

Test takers who do not achieve the minimum passing score may retake the exam by purchasing a new exam during their eligibility period. However, BCSP recommends that such individuals review their score reports and carefully study sections on which they did not do well, lest the score not improve substantially on a second or subsequent attempt. Reasonable accommodations for test takers with documented disabilities are available upon request through BCSP at the time an exam is purchased.

Test Format

The CHST exam is administered by computer. It is four hours long and consists of 200 multiple-choice questions with four possible answers for each question. Test takers may answer every question in sequence or they may skip questions, mark them, and readdress them later in the exam period. Use of any outside reference materials is prohibited; calculators on an approved list of models may be used. Test takers may take breaks whenever they wish, but the four-hour clock will continue to run.

The content of the exam is as follows:

Domain	Share	Description
1: Hazard Identification and Control	57.1%	Common hazards and controls including corrective actions, testing, and monitoring
2: Emergency Preparedness and Fire Prevention	10.3%	Emergency response, fire protection, and first aid
3: Safety Program Development and Implementation	17.1%	Health and safety standards and worksite assessment
4: Leadership, Communication, and Training	15.5%	Requirements and strategies for training, communication, and information management

Scoring

BCSP uses a panel of experts to rate questions with respect to a minimally qualified examinee. A statistical process incorporating these ratings is then used to determine a minimum passing score on the exam. Test takers who meet this passing score (and whose application for certification has been otherwise approved) receive CHST certification. As of 2016, the passing score for the CHST exam was 61.7%.

A test taker's score is based on the total number of correct answers achieved. There is no penalty for guessing, and all items are weighted equally. Test takers will receive their results before leaving the testing facility.

Hazard Identification and Control

Recognizing hazards in construction activities goes far beyond the most common types of hazards such as falls-from-height and confined-space hazards. Effective hazard assessment efforts must include an in-depth analysis of all operations. When construction activities include processes such as welding, demolition, use of hand tools for grinding and cutting, and use of blasting agents, the resulting environment becomes a matter of **industrial hygiene**, or the protection of workers from health and safety issues.

Like other hazards, industrial hygiene issues must be recognized and assessed prior to the beginning of work activities. Subsequent control measures can then be used to mitigate risk to workers and the environment. Risk related to chemicals, toxins, and airborne contaminants is often based on the amount (or concentration) of the substance present and the duration of exposure. Common environmental and industrial hygiene issues are classed into three categories. These categories, along with the processes and substances associated with each, are as follows:

Chemical Hazards	Physical Hazards	Biological Hazards
Lead	Noise	Mold
Asbestos	Vibration	Bird droppings
Welding	Radiation	Bacteria and viruses
Solvents	Ergonomics	Poisonous plants
Corrosives	Heat and cold exposures	Animal/insect bites

Many types of evaluation methods are used to determine the extent of exposures to such substances. Methods include real-time monitoring or obtaining a sample of a substance in the work environment where results are compared to regulatory limits. Results can be used to determine if controls are necessary or if applied controls are adequate.

For many of these hazardous substances, **OSHA permissible exposure limits** (PELs) outline how organizations must comply with regulations regarding exposure to these substances and allow OSHA to enforce the limits of use and punish noncompliance to regulations. In contrast, the limits and values set by other organizations are recommendations and are unenforceable. The **National Institute for Occupational Safety and Health** (NIOSH) lists **recommended exposure limits** (RELs). **Threshold limit values** (TLVs) are held by the American Council of Governmental Industrial Hygienists. Organizations can apply TLVs for a higher standard of worker protection, as they are often more stringent than the limits of OSHA.

Hot Work Hazards and Controls

Hot work is work that produces heat, fire, or sparks; examples include burning, cutting, welding, brazing, grinding, and soldering. This type of work presents many hazards from burning or electrical shock.

Hot work produces heat, sparks, or flame. Hot work presents the risk of burns from contact with hot tools or molten metal. Sparks can fly up to 35 feet and get caught on clothing, work surfaces, and unprotected openings in floors, walls, or construction material. Workers are at risk of being burned if flying sparks ignite when they come into contact with flammable materials on the work site such as wood, paper, soiled rags, clothing, dust, flammable liquid, or chemicals.

Hot work also presents the risk of workers being injured by explosions if flying sparks or hot tools come into contact with flammable gases that have leaked into the work space.

Heat-producing tools that run on electricity present the risk of electrical shock if the tool malfunctions.

To prevent serious injuries, workers should perform hot work only in a safe, well-ventilated location from which all combustibles have been removed. The work space should be enclosed in fire-resistant material, such as a welding curtain, to contain the heat and sparks. If workers cannot completely enclose the area, all combustibles should be protected by fire-resistant material such as welding blankets. To prevent explosions, the work space should be tested for flammable gas leaks before hot work is started.

To prevent electrical shock when using heat-producing tools, workers should make sure tools are properly grounded. Insulating mats should be used to prevent contact with fire and sparks.

Workers should always wear appropriate personal protective equipment for hot work. This includes welding gloves, eye and face protection, oil-free and fire-resistant clothing, and foot and leg protection such as tall leather boots with steel toes.

Electrical Hazards and Controls

OSHA standards require and assume that only qualified employees perform electrical work. A similar position is taken by the practices and procedures outlined in **NFPA 70E** (not an OSHA standard). Top electrical hazards include work near overhead power lines, electrical equipment and **lockout/tagout** (**LOTO**) issues (verifying machines are properly turned off at the end of each shift), wiring, and lighting fixtures. Providing measures of protection for electrical hazards are the responsibility of the employer. This begins with providing adequate training for employees that should consist of: evaluating hazardous job conditions, work processes, and precautionary items; energy isolation controls; approach distances for low-voltage (50-600 V) and high-voltage (> 600 V) panels; and appropriate PPE.

Control items include using double-insulated power tools, rubber coated non-conductive hand tools, and materials used near energized power lines and equipment. Gloves, boots, clothing and coverings, electrically-rated hard-hats, and other protective equipment must be inspected and meet their associated **ANSI** requirements. PPE selection is dependent upon several factors including voltage, current, distance, and duration of exposure. Testing meters must be approved and of an adequate voltage rating. A vital measure of protection includes the installation of **Ground Fault Circuit Interrupter (GFCI)** protection in wet areas and with all tools and equipment deemed temporary. Inspections of tools and equipment should be performed to detect hazards such as torn and damaged insulation, exposed wiring and junctions, shortened or altered cords, and circuit overload.

When employees work on (or service) machines and/or equipment, the appropriate energy control procedures must be used to prevent worker exposure to the release of hazardous energy. This involves de-energizing equipment, performing the correct sequential shutdown procedures, and properly detecting and isolating energy sources in conjunction with effectively using LOTO devices and equipment. In addition, the possibility of accumulating and releasing residual energy must be continually monitored throughout the service and repair process. During construction activities, common LOTO injuries and fatalities result from the hazardous energy released by concrete mixers. Isolation devices for machines and equipment include circuit breakers, in-line ball and gate valves, disconnect switches, chains, etc.

Lockout/Tagout equipment consists of padlocks and keys, locking safety hasps, blocking devices, guards, covers, and tags with warning labels stating: "Do Not Operate" (or a similar wording). Lockout materials must be in good condition and be:

- **Durable**: Materials fit for the conditions
- **Standardized**: By size and shape
- **Substantial**: To withstand stresses and prevent removal
- **Identifiable**: For visual continuity

The employer is responsible for an **Energy Control Program** that must include the following:

- Written energy control procedures
- Employee training
- Periodic inspection
- Enforcement and penalization with documentation
- Annual inspection

LOTO and energy isolation procedures are designed specifically for the machine or piece of equipment. The generalized lockout sequence is as follows:

1. Prepare the machine for shutdown and notify the affected personnel.
2. Shutdown the machine or equipment.
3. Identify and isolate the energy source(s).
4. Apply LOTO devices.
5. Release or relieve stored energy and monitor residual accumulation.
6. Try out the machine – for verification of procedure effectiveness.

Electrical Cord Safety

A lack of electrical cord etiquette is a common housekeeping construction site deficiency. Electrical cords are always present and must be dealt with so that trip hazards are eliminated. This can be as simple as picking up electrically powered tools instead of leaving them where they were last used until they will be needed again, which could be a few hours or later in the week.

Sloppiness has a tendency to spread. If employees leave their tools on the ground where others could trip on them or their cords and this hazard is not corrected, other employees will follow suit because it will be perceived as an acceptable practice.

Just one tool with a long extension cord can cross over a common travel path in multiple places, each one of them a potential trip-and-fall scenario. This oversight is especially problematic on stairways. Most falls on stairs occur while descending; a loss of balance, momentum, and gravity are a hazardous combination. Adding a potential trip hazard ensures a fall is likely to occur.

Cords can be routed along walls to keep them off the ground, but they should not be strung up by staples or nails. This is a very common practice on construction sites, but it should be avoided, as it has resulted in death. While rare, it has happened and is the reason why OSHA prohibits nails, staples, and all other sharp objects and surfaces from being allowed to contact electrical cords.

The amount of current generally considered to be fatal is 70 to 100 milliamps, although fatalities have occurred at lesser amperages. Of note, 1 milliamp is 1/1000 of 1 amp; 100 milliamps is 1/10 of 1 amp. Most circular saws draw between 12 and 15 amps; this equates to 12,000 to 15,000 milliamps, more than one hundred times the current required to instantly kill. Most drills draw between 5 and 7 amps; again, this is far more current than the lethal range. On a construction site, nearly all electrically powered tools draw sufficient current to be fatal if contacted.

When removing the cord from the wall, the cord could catch on the nail or staple and shear off a chunk of insulation, which would expose the wiring. This is more likely if the cord is pulled from the wall. An employee could inadvertently handle the cord where the wiring is exposed; an additional risk may occur when the hazard goes unnoticed.

Another practice that should be discouraged is routing electrical cords through holes in walls or floors. Since workers cannot see what is inside the wall or floor or what is on the other side, they cannot know if the insulation is at risk of being sheared off from inadvertent contact with a sharp object (for example, a piece of metal flashing or a pile of sheet metal shards).

When run through doorways, the cord must be protected against damage. If a door closes on an electrical cord, it is unlikely to sever it, and the door usually provides insufficient force to significantly damage the insulation, although damage is possible. Pinching an electrical cord with a door could damage and fatigue the wiring, which would not be obvious when the insulation is not damaged. Although a wire may not be severed, it will begin to break when the cord is handled because a pinch will crimp the wire. A cord will bend at the crimp point and fatigue to the point of breaking over time. When there are fewer and fewer connected wires in the cord to draw current through, the wires in the cord will overheat and present a fire hazard.

When electrical cords are damaged, they should be removed from service immediately. Any break in the insulation, separation of the plug from the insulating sheath, or broken or damaged prongs should disqualify the cord from continued use.

A damaged cord can be replaced, or it can be repaired. If repaired, it must be repaired to a factory state. The cord doesn't have to be repaired by a licensed electrician, but it must be repaired in such a professional manner that it is indistinguishable from an undamaged cord that could be purchased new. Using electrical tape to repair insulation or any other "quick fixes" should be avoided.

Continuous Path to Ground

The reason to have a continuous path to ground is to reduce the likelihood of current passing through a body if it somehow escapes the protection afforded by insulation. The ground prong on a power cord helps the current to get to ground as efficiently as possible and without employees being seriously harmed in the process. For this reason, a ground prong should never be removed from an electrical cord. Removing the ground prong will eliminate a component of the power cord intended to make it more likely that a worker will survive an electrical incident. However, this is a fairly common practice. OSHA frequently cites this violation of its electrical safety standards.

Grounding tools and equipment during construction is mandatory with one exception: double-insulated tools. Instead of a ground prong (third wire) on the power cord, a double-insulated tool will have an extra layer of insulation, since the insulation alone is the only protection available. Double-insulated tools are indicated by a symbol: a square inside a square. They must have this symbol on them; it is the only way to readily identify a tool with no ground prong as being double insulated.

Older tools that are neither grounded nor double insulated that are encountered on construction sites should be removed from service immediately. OSHA allows no exemptions for older power tools or equipment manufactured prior to the enactment of the 1926 standards in 1971; they cannot be "grandfathered" in.

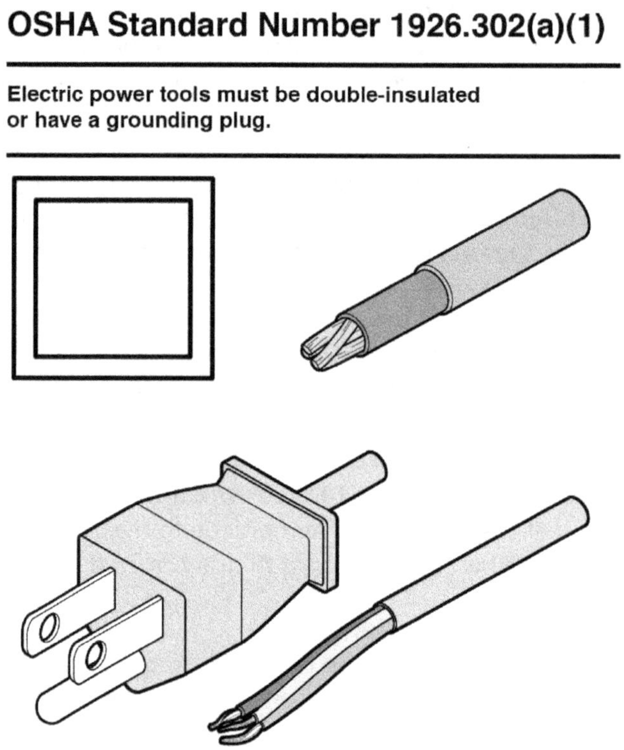

Ground-Fault Current Interruption (GFCI)

OSHA's 1910 General Industry Standards require that GFCI be provided for all installations in bathrooms and on rooftops, but in construction activity, GFCI is required everywhere with one exception. The only alternative to GFCI in construction activity is by using a formal "assured equipment grounding conductor program." However, this is not a common practice and should not be considered a viable alternative to GFCI in most cases. An assured equipment grounding conductor program requires a substantial amount of documentation and equipment testing that, in most cases, will cost much more than the potential savings of foregoing the comprehensive use of GFCI-protected devices on a construction site.

A GFCI-protected electrical device will interrupt current when the device detects a disparity between the hot and neutral return wires. The amount of current in the wiring should always be the same in both directions; when it is not, a significant electrical hazard exists.

The difference in current between the two wires is not lost. Instead, the excess current goes to ground. When functioning properly, a GFCI device will shut off current when the difference in the two wires is between 5 and 7 milliamps. The interruption will take place within 1/40 of a second (0.025 second). GFCI devices should be inspected and tested frequently to ensure they are functioning as intended; GFCI testers can be readily acquired in hardware stores and large retailers.

A common misconception is that GFCI does not provide protection for non-grounded double-insulated tools because they do not have a ground prong. This is incorrect because the ground wire plays no part in the function of the GFCI circuitry; rather, the hot and neutral return wires are affected, so double-insulted tools will work with GFCI devices.

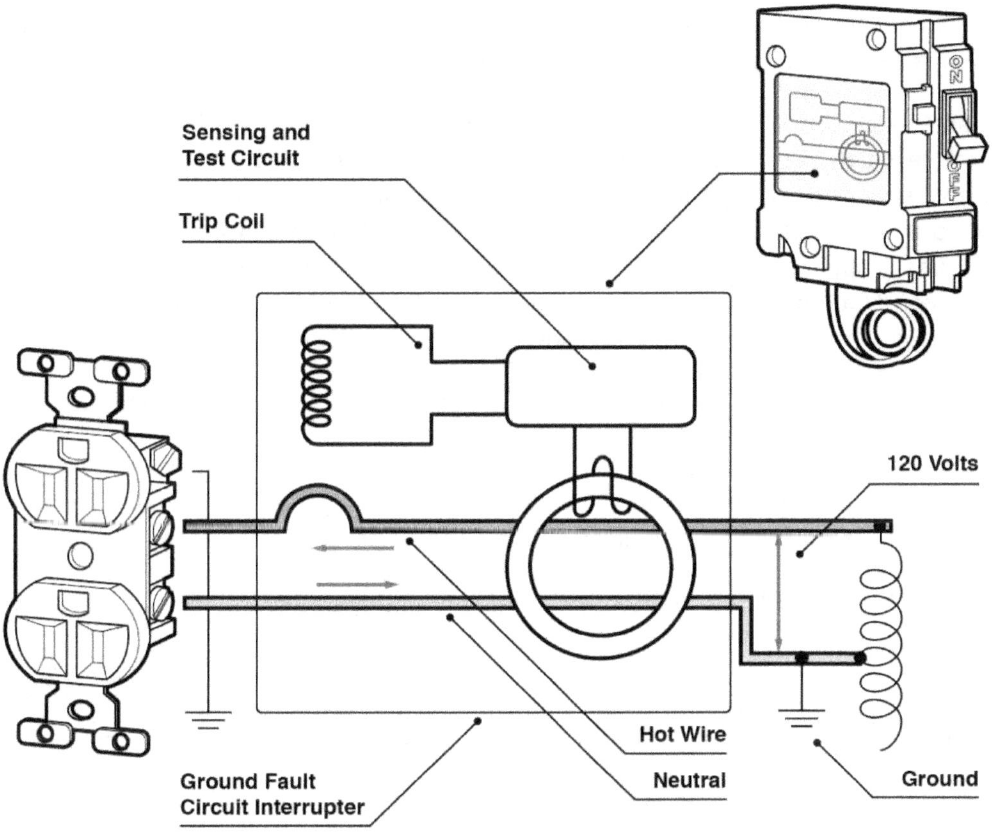

A GFCI-protected electrical device will interrupt current when the device detects a disparity between the hot and neutral return wires.

The type of GFCI installation is up to the contractor, who can choose to have every single receptacle equipped with GFCI, use a converter box, or employ a centralized circuitry system that will provide GFCI protection to all electrical tools plugged into the system. As with all electrical devices, the system employed must be inspected and tested to ensure safe operation and should be removed from service when deficiencies are identified.

Excavation Hazards and Controls

Many construction projects begin with the trenching and excavation of the site. The process is performed within varying soil types. Knowledge of soil type and classification are vital for the protection of employees who work in construction trenches. A soil classification system with guidelines and regulations (and how they pertain to soil type, trenches, and excavations) exists in **OSHA 29 CFR 1926 Subpart P.** Trench and excavation hazards include crushing, suffocation, struck by, and falls. The proper classification of any soil type pertaining to the excavation/trench, in conjunction with the application of OSHA's 1926 Subpart P, provides important information for the identification, analysis, and remediation methods that minimize these hazards.

The mechanics of a soil/trench collapse begin with the factors of soils' heavy weight per cubic foot, gravity, and compounding factors with increased trench depth and the presence of water. Stronger soils are more dense and self-supportive, while weaker soils are less dense and more granulated. Depending on soil type and saturation, the lateral pressure per foot squared increases to the point where the soil or trench can no longer support itself. These factors determine the necessity for shoring and shielding, and the requirements for angle of slope and benching techniques.

Soil analysis and classification are performed using various methods and devices. Soils are classified into four types with respective compressive strength ratings in **tons per square foot (tsf).** These four soil types are: **stable rock, type A, type B, and type C**. When classifying soils, multiple testing methods should be performed on soil samples. Results are analyzed and recorded for determining safeguards and design techniques. Sloping and benching methods correspond to soil type. Shoring and shielding equipment are constructed of primarily aluminum and timber. Methods, terms, and devices associated with trench and excavation activities are as follows:

- **Shield**: Single unit with bracing designed to protect workers in the structure from trench collapse/cave-in.
- **Shoring**: Multiple units that protect workers by supporting the sidewalls of trenches and excavations. Shorings are designed to create a longer area of protection.
- **Stable Rock**: Solid and non-fissured rock that can be sloped vertically or 90 degrees from horizontal.
- **Type A Soil**: Cohesive soil with a tsf rating of 1.5 or greater. Designed slope can be ¾ to 1 (rise over run percentage) or up to 53 degrees from horizontal and can be benched.
- **Type B Soil**: Cohesive soil with a tsf rating between 0.5 and 1.5. Dry but determined unstable due to fissuring or being previously disturbed. Designed slope can be 1 to 1 or 45 degrees and can also be benched.
- **Type C Soil**: Cohesive soil with a tsf rating of 0.5 or less. Soil is granulated, saturated, or submerged. Designed slope can be 1.5 to 1 or no greater than 34 degrees. Type C soil cannot be benched.
- **Visual Analysis**: A visual examination method of viewing and inspecting the soil sample for cohesiveness, fissuring, layering, and granular composition.
- **Plasticity**: Measured by molding the soil sample into a small ball, then rolling it out to a minimum thickness of 1/8 inch. Cohesiveness/Non-cohesiveness is determined by holding a minimum 2-inch-long piece at one end to see if it tears apart under its own weight.
- **Dry Strength**: Determined by examining the sample behavior when crumbled by varying degrees of hand pressure. Sample reaction reveals the ratio of clay to gravel, sand, or silt.
- **Thumb Penetration**: Method of applying thumb pressure to an unconfined and cohesive soil sample. This reveals an approximation of a tsf rating applicable to soil types A, B, and C. A high degree of penetration difficulty corresponds to soil type A; a moderate degree corresponds to type B; and penetration with light pressure and minimal effort corresponds to type C.

- **Pocket Penetrometer**: A penetrometer contains a spring-loaded pin, gauge, and a stop ring on the body of the device. The pin on the end of this device is pushed into a section of a soil sample until the soil contacts the edge of the ring/stop. The penetrometer measures unconfined samples of compressive strength in tsf.

- **Tor/Shear Vane**: This method is performed on undisturbed or "in-form" soil by depressing the fins of the instrument into the soil sample until the soil contacts the backing plate behind the fins. Zero calibration ensures measurement accuracy and is done before twisting the device until the embedded portion (fins) shears soil away from the section. This instrument also measures in tsf.

The presence of water in soil and standing water exacerbate the possibility of a trench collapsing. Standing water must be removed from any trench before worker entry. In accordance with OSHA guidelines (1926 Subpart P App. B), a registered engineer must design all excavations to be deeper than 20 feet. Site programs must consider additional hazards including: the weight of nearby equipment and associated vibration; fall hazards greater than four feet; oxygen deficiency; accumulation of toxic fumes within trenches from other equipment and processes; rescue plans; and a means of egress where the placement of ladders, ramps, etc. must accommodate workers' lateral travel to 25 feet or less.

Hazards and Controls Associated with Heights

A fall from height occurs when a worker loses balance and becomes subject to the very force keeping humans on the planet: gravity. Falls from height are one of the leading causes of death in the construction industry alongside being struck by an object, caught in or between objects, and electrical hazards. **Fall hazards** are worksite conditions or materials that may cause a fall and include falls from ladders, scaffolding, structural steel, and roofs. They result from unprotected sides on raised platforms such as roofs, scaffolding, or uncovered floor or roof openings (others may include unsecured ladders or scaffolding in poor condition). **Same-level falls** are generally caused by trip hazards and poor worksite housekeeping that result in neglected ground obstacles or impediments (one of the leading causes of injury in the construction industry).

Fall Hazard Sign

Fall protection is the measure or measures taken to prevent individuals from falling. OSHA dictates fall protection is applicable to work done above four feet (general industry), five feet (marine), six feet (construction), and specifically, ten feet for scaffolding.

Duty to Provide Fall Protection

OSHA's requirements regarding fall protection can be found in 1926, subpart M. For most working surfaces in construction, such as roofs, edges, and floor openings, fall protection must be used six or more feet above the ground or the next lower working surface. Fall protection is also required when a fall is possible onto dangerous equipment, machines, or materials when less than six feet in height.

OSHA's general rule to provide fall protection at 6 feet differs in some situations in construction. Fall protection is required on scaffolds 10 feet above the ground or the next lower working surface. Scaffold safety provisions can be found in 1926, subpart L. Fall protection on aerial lifts is always required, regardless of height. Although treated differently than scaffolds, OSHA's aerial lift requirements can also be found in 1926, subpart L. In some circumstances involving steel erection, fall protection must be provided 15 feet or more above the ground or the next lower working surface (20 feet in some scenarios.) These requirements can be found in 1926, subpart R. Fall protection guidelines for stairways and ladders can be found in 1926, subpart X.

Fall Restraint vs. Fall Arrest

The majority of fall protection systems in construction can be categorized as either fall restraint or fall arrest. Fall restraint includes guardrails and also harness devices that will not allow a fall from height to occur. With a fall arrest system, such as a **personal fall arrest system** (PFAS) or safety nets, a fall can happen, but a device stops the fall.

While contractors are expected to select the system that provides the most reliable means of protecting employees from fall hazards, OSHA generally prefers restraint systems to arrest systems because a fall does not occur at all when a restraint system is installed, such as a standard guardrail system. Fall restraint systems also tend to be less complex than arrest systems and so do not require the level of advanced technical proficiency needed to effectively employ a fall arrest system.

Guardrails

Guardrails are a fall restraint system that is most frequently installed on flat working surfaces. Although technically allowed in some circumstances, the use of guardrails on sloped surfaces is not advised for various reasons, most importantly because, even when strong enough to prevent a falling employee from breaking through, the employee could be hurt when contacting the guardrails.

OSHA's requirements for guardrails in construction, found in 1926, subpart M, specify that the top rail be installed at 42 inches from the working surface, plus or minus 3 inches; so, anything between 39 and 45 inches is acceptable.

When a midrail is installed, it must be approximately half the height of the top rail. Midrails are not mandatory but are very common. An alternative to a midrail is to install a wire mesh or other suitably strong material from the top rail down to the working surface. Vertical supports (balusters) may also be used instead of a midrail but must not be farther than 19 inches apart.

A guardrail system must be strong enough to stop a fall; OSHA specifies that the top rail be able to withstand a force of at least 200 pounds (890 Newtons). While caution tape may allow some notification of the proximity of a roof edge on account of its highly visible color, it should never be permitted to serve as a guardrail system because it will not stop a fall.

Personal Fall Arrest System (PFAS)

PFAS setups require a lot of planning and testing to ensure they will work as intended. In contrast with guardrail systems, a PFAS is much more complex in every regard and should be used only when guardrails are not a feasible means of fall protection (for example, on sloped working surfaces such as pitched roofs).

A PFAS includes a full-body harness (body belts are not acceptable), lanyard, other devices and connectors, such as D-rings, and the anchorage point itself. All of these components must meet strength requirements and be tested to ensure safe function. Five thousand is a good number to remember with PFAS.

An anchorage must support 5000 pounds per worker attached or a safety factor of two. Since the latter involves additional testing with a representative weight and desired drop distance to ensure it will be effective in a given circumstance, most fall protection specialists will select an anchorage point verified to support 5000 pounds per worker. Note that 5000 pounds is not referring to total weight but the weight in force that will be transmitted to the anchorage point when the fall is arrested.

Lanyards and vertical lifelines must have at least 5000 pounds of breaking strength. D-rings and snap hooks must have at least 5000 pounds of tensile strength.

A CHST must understand that, even when functioning properly, PFAS can still injure an employee when the fall is arrested, and a fall restraint system should be used whenever practical to do so.

To minimize the likelihood of a fall arrest causing an injury, OSHA permits a drop of no more than six feet. The maximum fall distance allowed is based on the fall distance itself, not the length of the lanyard alone, so the selection of the anchorage point and all components must be chosen carefully to ensure that an employee cannot fall more than six feet. Shock-absorbing lanyards will reduce the likelihood of an injury during a fall arrest because they reduce the force transmitted to the employee's body. Shock-absorbing units destroy themselves when an arrest occurs by tearing (stretching.) Since they will lengthen when arresting a fall, this must be taken into consideration when planning to keep the maximum fall distance to six feet. Shock-absorbing lanyards can never be used after they have been stretched in a fall arrest situation.

Scaffolds

Scaffolding is a work platform that can be supported from the ground or suspended. This equipment is used to support workers and associated materials during construction activities. The **Occupational Safety and Health Administration** (OSHA) provides standards for 25 types of scaffolding. Ladder jacks, pump jacks, aerial platforms, and aerial lifts are also considered scaffolds. For equipment to be recognized as a scaffold, it must be:

- Temporary
- Elevated
- Designed to safely support both workers and material

The most common type of scaffold is **tubular welded frame**, which is assembled from prefabricated framing and cross braces. Walking and working areas should be equipped with scaffold-grade lumber. Decks must be fully planked where work is performed, while areas exclusively used for walking and traversing require a minimum planking width of 18 inches. Scaffold parts must be of the same type with no mixing of components. Supported systems must have a load capacity of four times the intended load, while suspended systems must support six times the intended load. Scaffolds require fall protection beginning at a height of 10 feet without the presence of guardrails. As a general rule, scaffold units exceeding a height-to-width ratio greater than 4:1 must be tied in. Tie-ins are also required at width intervals of 30 feet. Scaffolds should never be climbed upon and they require safe points of access with the presence of ladders. Supported units on asphalt or soft ground require baseplates coupled with mudsills to prevent sinking, twisting, or potential collapse. A competent person using a tagging system should inspect scaffolds before each shift.

Scissor lifts are commonly encountered in construction. Most scissor lifts in use today are a type of mobile scaffold; scissor lifts that require movement and placement to be accomplished with another device are not common in the construction industry, although they do exist and have their purpose. If a scissor lift is mobile, then OSHA treats it as a mobile scaffold, and applicable safety requirements for scaffolds will apply.

Most of the general safety principles that apply to fixed scaffolds will also apply to scissor lifts, for example, strength and suitability, proper footing, sufficient working space on platforms, and the common use of guardrails. As with nonmobile scaffolds, a PFAS is not mandatory when working on an elevated scissor lift platform. Some scissor lifts do have an anchorage point on the deck for attaching a PFAS, but this is not very common, since PFAS is not mandatory as is the case with aerial lifts. A guardrail system is the predominant means of fall protection employed on most scissor lifts.

Additional precautions are necessary for powered and mobile scissor lifts. When moving a scissor lift, employees should be permitted to remain on the platform only when it will be moved slowly and safely, when the traveled surface is within three degrees of level, and when there are no holes or obstructions that could prevent safe movement. The scissor lift should be lowered when traveling to reduce tipping. OSHA requires the scissor lift to be no higher than twice the width of the base when traveling with mounted employees; the only exception is when the device's engineering parameters are such that it will remain stable when traveling with a height to base width ratio exceeding 2:1.

Mobile scissor lifts are sometimes misidentified as aerial lifts. While both types of equipment involve elevated working platforms, the methods of fall protection permitted by OSHA's 1926 standards are significantly different. An aerial lift, as defined by OSHA, is easily identified by a boom that is extensible (telescoping) and/or articulating (multiple planes of movement). A scissor lift will not have any components that telescope or articulate; it will only go up and down on its scissor mechanism.

A CHST should understand that, while OSHA's guidelines for aerial lifts can be found in a section contained within 1926, subpart L, scaffolds and aerial lifts are not all treated as scaffolds, and general scaffold fall protection requirements do not apply to aerial lifts.

Unlike scissor lifts and other scaffolds where guardrails are the norm, OSHA requires that a harness be used whenever working from an aerial lift, even when guardrails are installed. The requirement to wear a harness applies regardless of the height of the platform, even when the platform is less than six feet from the ground. A PFAS utilizing a full-body harness must be used when a fall from the aerial lift platform is possible; this is the same treatment as when working on any other surface when guardrails are not feasible.

A CHST should advise that the PFAS be attached to an anchorage point on the deck, and not the boom, railing, or anything else higher up whenever possible. If this is done, then the total fall distance prior to the arrest will be minimized in the event of a fall from the platform. OSHA allows up to a 6-foot drop in a

PFAS, although this should always be minimized, since an arrested fall can still cause back or other injuries.

Assume the total length of the PFAS components (not just the lanyard) will arrest a fall at six feet exactly. If the PFAS is attached to the boom or the top of the guardrail or bucket, then the fall arrest distance will be approximately six feet. However, if the anchorage is on the deck, then the fall arrest distance is reduced because the fall will be limited to the length of material that goes beyond the top of the platform and not its total length. Minimizing the fall arrest distance will also help to minimize the likelihood of an injury resulting from an arrested fall.

If the anchorage point and chosen PFAS components are such that an employee cannot physically go over the edge, then there is no fall arrest situation at all. In such situations, OSHA even allows the use of body belts, and not a full-body harness, since there is no fall potential. OSHA permits the use of a **personal fall restraint system** (PFRS) utilizing a body belt when a fall is absolutely not possible, although this practice has mostly become unpopular on aerial lifts, and a CHST should advise the use of a full-body harness even when a restraint system would be in compliance with OSHA's standards. Another consideration when introducing body belts for use solely in restraint situations is that employees may see them used and incorrectly assume that they are acceptable in fall arrest situations and utilize them as such.

There is one instance when not wearing a harness in an aerial lift is allowed and may well be advisable. This is when working over water and the drowning hazard is greater than the fall hazard. When drowning is deemed the greater hazard, a CHST should ensure that employees are wearing an approved personal flotation device and that means are available to get them out of the water as soon as possible such as a small boat. When correcting hazards in construction, a CHST must avoid the introduction of even greater hazards in the process.

Stairs and Ladders

Construction activities sometimes require the use of stairs and ladders; it may be necessary to have stairs built for the crew to use during the building project. OSHA guidelines (1926.1052(a)(2)) state that stair construction requires a uniform riser height and a tread depth of no more than a ¼ inch variation of each element. The stair slope or angle must fall between 30 and 50 degrees. Stairs with the presence of four or more risers (or exceeding 30 inches in height) require at least one handrail that's a minimum of 36 inches in height above each stair nosing. The handrail must be capable of withstanding at least 200 pounds of side force, and a midrail must be present halfway between the handrail and the steps.

Ladders should be set using a 4:1 ratio where the point of contact at height distance is four times greater than the point of contact distance at the base (away from the wall). When in position and secure, ladders must extend three feet past the surface of the upper landing. Safe use requires the worker to face the ladder at all times, use at least one hand to grasp the ladder (three points of contact), avoid carrying any load that would cause loss of balance, and never use or perform work on the top rung. OSHA requires fall protection at a height at above 24 feet for fixed ladders.

Workers who climb less than 24 feet in height on a fixed ladder that still extends up to at least 24 feet also need to have cages, ladder safety devices, wells, or self-retracting lifelines provided. Extension ladders less than 36 feet should maintain a minimum overlap of three feet; those between 36 and 48 feet should maintain a minimum overlap of four feet. Ladders between 48 and 60 feet require a minimum overlap of five feet. A competent person should inspect step ladders and extension ladders before each use. When a ladder is removed from service, subsequent repairs should be done to return the ladder to its original design criteria.

Hazards and Controls Associated with Confined Spaces

Confined space (CS) activities present a multitude of hazards that can result in unnecessary fatalities. Most fatalities result from air quality issues and hazardous atmospheres. Of all confined space fatalities, the overwhelming majority have traditionally been supervisors, project managers, and would-be rescuers attempting to save a coworker or relative. Nearly 25 percent of fatalities are from entry into a space that already contains deadly conditions. In addition to hazardous atmospheres, confined space hazards include drowning, engulfment and suffocation, blunt force trauma, entanglement, electrocution, and lockout/tagout issues that contribute to these hazards. A confined space must have all three of the following characteristics:

1. The space must be large enough for a person to fully enter and perform work.
2. The space or area cannot be intended for continuous occupancy.
3. Any means of entry and exit must be limited.

Examples of confined spaces include tanks, manholes, vaults, trenches, and ventilation ducts. Some confined space activities require a permit before work can be performed. Permits for confined spaces require the recording of general information, hazard evaluation, initial air testing, entry preparations, communications, air test records, and the supervisor's signature. Confined space permits must be kept on file for one year. For a confined space to require a permit, it must have at least one of the following conditions:

- A structure that can trap a worker through falling walls or tapered floors, or is inward converging or tapering
- Hold or has the potential to acquire hazardous atmospheres
- Hold material that has the potential to engulf a worker
- Contain other detected hazards and lockout/tagout issues

Testing and monitoring are performed for the presence of gases and other potentially hazardous or explosive atmospheric constituents. These are done to detect the presence of carbon monoxide, inherent or introduced toxins and gases, and oxygen deficiency and enrichment, which can impact hazardous atmospheres including the **lower explosive limit** (LEL) and **upper explosive limit** (UEL). It is vital for the safety professional to know and consider factors of vapor density and the link between oxygen levels and potential modifying effects on LEL/UEL. Monitors are set with a safety factor so that an alarm is triggered if oxygen levels fall outside the range of 19.5 to 23.5 percent in atmosphere. They are also set at 10 percent and above for the LEL with respect to the presence of flammable gases, fuel-air mixing, combustion, and explosion.

Control methods also include the ventilating, purging, and inerting of hazardous atmospheres. Purging and ventilating include the use of air and/or water to flush hazardous atmospheres. Ventilating can be done via supply, exhaust, and the use of positive and negative air pressure. Inerting involves the use of nonhazardous gases, such as nitrogen and argon to push, to expel hazardous atmospheres from confined spaces.

Confined space operations must have a minimum of two workers: the **entrant** and the **attendant**. Entry supervisors authorize and oversee the operation's elements and entry. With proper training, the supervisor can serve as the attendant or authorized entrant. Supervisors must test and review the feasibility of retrieval systems (e.g., tripod/rescue harnesses for non-entry rescues) to determine system effectiveness and the potential for creating additional hazards like entrapment or entanglement.

Attendants monitor the entrant(s) and the conditions associated with confined space activities. When tracking the number of workers in a confined space, attendants must be knowledgeable of atmospheric hazards, and must monitor means, equipment, and worker symptoms while maintaining constant communication. The attendant must also monitor the activities outside of the confined space and keep unauthorized personnel away from the area. They must never perform other tasks that would interfere with these duties, and they must never enter the confined space unless relieved by another qualified attendant.

Struck-By Hazards and Controls

This term is defined solely by an impact on a worker by equipment or an object. There are four categories described for **struck-by** events:

- **Struck by a flying object**: Generally, a horizontal event caused by energized objects propelled or hurled across a space (typically associated with pressure or compression).
- **Struck by a falling object**: Generally, a vertical event caused by falling objects or equipment (suspended loads, unsecured tools, or equipment falling from height).
- **Struck by a rolling object**: Generally, a horizontal event that may include unsecured equipment such as pipes or wheeled equipment and may include vehicles or mobile equipment.
- **Struck by a swinging object**: Generally, a pendulum or slippage event associated primarily with crane work, poorly rigged loads or crane malfunction, crane operator error, or a worker mistakenly in a restricted area. It is imperative that the CHST marks and communicates the area where crane-lifting operations are in process.

Awareness of the environment on all three planes (ground, line of sight, and overhead) will help to keep workers safe. The CHST must keep a sharp eye out when (1) cranes are performing lift operations or changing stations; (2) scaffolding is in use, being erected, or disassembled; and (3) high build work or activities are being performed in the vicinity of a marginal structure.

Caught-In or -Between Hazards and Controls

As stated by OSHA, caught-in/between hazards consist of a worker being caught, pinched, compressed, squeezed, or crushed between parts of an object (machinery) or by two or more objects (mobile and/or stationary and vehicular). The CHST should know the difference between a struck-by and a caught-between hazard: A struck-by hazard is when only the impact from an object causes injury, and a caught-in/between hazard involves a worker being crushed between objects. Events that may be classified as caught-between include being pulled into or caught by machinery, pinned in between an object and structure (vehicle and building) or between two or more objects (energized machinery and/or equipment), and cave-ins (during trenching, excavation, or demolition work).

The construction industry accounts for approximately one quarter of all private industry caught-in/between events, and these events constitute about 10 to 12 percent of overall hazard events annually. Mitigation measures for cave-ins during trenching and excavation work have successfully reduced accidents in this category. It is the onus of the employer to provide the training, equipment, and environment mandated by OSHA for the identification and prevention of caught-in or caught-between events on the worksite.

As with *any* hazard, or any incident that occurs as the result of a hazard, **corrective and preventive actions** (CAPAs) must be implemented to mitigate and minimize, if not eliminate, future threat. The CHST, along with other project management personnel, is responsible for helping to identify the root cause of each identified hazard or incident. From there, a corrective action plan must be in place and, moreover, implemented in a timely manner, in order to address each hazard or incident effectively. In addition, preventive actions should be planned and implemented to address future hazards and possible incidents. Each plan should outline specific, assigned action steps and contain a deadline for completion. CAPA audits should be performed in follow-up to ensure all actions have been thoroughly addressed. Any critical, real-time hazard or incident that needs immediate corrective action must be fully documented after the fact to maintain compliance with international, federal, and state laws and regulations. *The CAPA process applies to all hazard types and all incident occurrences and must be fully implemented throughout the construction project in order to avoid costly human and financial loss.*

Hoisting and Rigging Hazards and Controls

Construction work requires large loads to be transported from one location to another. When loads are too large for workers to move manually, they lift them with hoists such as cranes. Rigging is the process of using slings to hoist loads. The slings are made of such materials as wire rope, metal chains, or synthetic straps.

Falls can occur when workers perform work on elevated surfaces or in the path of obstacles. To communicate with the crane operator to direct the load, workers sometimes must walk backwards on uneven surfaces or around other heavy equipment.

Workers can be struck by swinging material. For example, the blow can be caused by a crane operator losing control of their load, or the worker not paying attention to their surroundings. Workers can also be crushed by heavy loads that slip during hoisting due to being improperly secured. Workers can also get caught between loads and other objects on the work site such as heavy equipment, walls, or temporary structures.

The work site should be equipped with fall protection such as guardrails or safety belts. Before beginning work, workers should walk the work area to ensure they are aware of any uneven or slippery surfaces. They should also use the opportunity to clear any obstructions they might trip over. Workers should always wear personal protective equipment such as hard hats, face shields, and steel-toed boots, and they should remain out of the path of hoisted loads.

Crane Operation Hazards and Controls

Overhead cranes are used to hoist and transport heavy loads on construction sites. Like the rigging and hoisting process, crane operation presents hazards from improperly secured materials slipping, falling, or swinging out of control. Crane operation also presents electrical hazards and hazards associated with overloading the equipment.

During crane operations, heavy loads are hoisted over a work area where workers are usually present. If the loads are not properly secured during rigging, they can slip, and materials can strike workers or cause property damage. Workers can be pinched, crushed, or caught between swinging loads if crane operators are visually impaired or incompetent, or if the operators fail to pay attention to their work.

In addition to risk of shock from electrical tools, parts of the crane such as the boom can come into contact with active power lines and cause injury or death to the operator or to workers in the vicinity.

Exceeding the operational capacity of a crane can result in structural damage to the equipment, which can result in malfunctions that can cause serious injury. If the crane is overloaded, the load might slip and fall, causing injury to workers below.

To prevent loads from slipping and crashing to the ground, workers should be trained to properly secure loads and remain within the operational capacity of the equipment. To prevent electrical shocks, crane operators should be aware of power lines within 10 feet of operations. Workers should always wear protective equipment on the work site.

Material Handling Hazards and Controls

Material handling incidents constitute almost a third of construction workers' compensation claims and a full quarter of all industry claims. The sobering number and frequency of incidents speak to a common denominator between the similarity of these injuries and the repetitive nature of these occurrences. In short, most of these are known and can be mitigated, if not avoided altogether.

Work-related musculoskeletal disorders (WMSD) typically result from the repetitive and task-specific movements associated with construction and other heavy industries. Where injuries like sprains and minor lacerations would have been accepted in the past as an inherent part of the workplace, effort and attention are being directed to reduce the level of worksite incidents to **As Low as Reasonably Practicable** (ALARP) levels.

Alternative task methods and tool modifications have helped to alleviate the manual handling of heavier materials and provide improved ergonomics that prevent bodily positions typically associated with situational or chronic injuries. The extended rebar tie tool, the kneeling creeper, and the extended screw gun are examples of more recent tool modifications designed to alleviate potential hazards.

Manual material handling on construction sites continues to be a challenge. In some circumstances, limited working areas created by traffic zones, equipment obstruction, or confined spaces constrict worker movement and maneuvering space; this in turn hinders the capacity to use a neutral body position or mechanical aids for pushing, lifting, pulling, or carrying materials.

Although many mechanical aids for labor have been devised and employed at the worksite, lifting is still a mainly manual activity that is frequently performed. Poor lifting habits are a workplace staple. Workers will naturally lift an object in a manner that is most convenient, which precludes the proper body placement for optimum lifting (humans have spines, and cranes have booms; load capacity is reduced, and strain increases in both as they extend horizontally [boom out], torque, and deviate from a straight-up position). Proper lifting will prevent spinal column injury by avoiding lifts that are away from the body and that involve a twist while lifting. Children naturally lift properly. Their lack of musculoskeletal development forces them to squat down and position the load close to the spinal column before they lift. This, as it turns out, happens to be the proper lifting technique.

Standard recommendations regarding proper lifting include the following:

- When lifting a load, it should be kept close to the body (within ten inches).
- Do not twist or torque the body during lifting an object or setting it down.
- Keep the back straight, and lift with the legs.
- Use a solid, two-hand grip.

Injuries resulting from forces on the neck and spine can be greatly reduced by mechanizing the material handling process. Appropriated forklifts and carts can transport material from trailers or stockpiles to areas where the items are required. Some are even equipped to lift materials to required heights. Such equipment can reduce slip, trip, and fall injuries by reducing or eliminating carrying distances and traversing over uneven or slippery ground surfaces. Conveyor systems also can be added to the process, reducing the workload associated with manual tasks and minimizing risk.

With or without a standard hand truck (**dolly**), lifting and carrying can place heavy loads on the shoulders and spine. Powered dollies are an improvement over traditional dollies with the removal of awkward postures and force requirements. This equipment also is available with stair climbing features and adjustable handles.

Delivery trucks can be equipped with hydraulic lift gates, which are designed for loading and unloading materials and equipment. This eliminates the hazards associated with workers climbing on and off the rear of the truck, as well as the handling of other loading equipment.

Adjustable scaffolding provides continuous height adjustment. This allows brick and block layers to work from waist level, which minimizes back pain stemming from postures associated with bending and repetition. This system also enables planking to stay in place, eliminating the need to change it at different working levels.

Material Storage Hazards and Controls

The materials stored on a construction site pose both health hazards and physical hazards to workers.

Hazards associated with materials storage include the risk of contact with chemicals. Toxic agents can damage skin, eyes, lungs, and mucous membranes. Workers can come into contact with gases, dust, fumes, smoke, or vapors produced by these chemicals during normal use, handling, and storage.

Physical hazards are posed by handling and storing materials on a construction site. Risky activities include lifting and carrying loads, and there is a danger of being struck by falling objects. Potential injuries include sprains, strains, fractures, cuts, and bruises.

Some of the most common injuries on a work site are back injuries like strains or sprains caused by lifting loads improperly or by carrying excessively large or heavy loads.

Stored material can slip from high shelves due to being improperly stacked or overloaded, and workers can be struck and injured by the falling objects. Workers can also be struck if they lose control of or drop a heavy load they are carrying.

Workers might trip and fall if they cannot see obstructions over or around oversized loads.

Workers should wear personal protective equipment when handling chemicals on the job site. They should also follow the instructions for safe handling on the Material Safety Data Sheets. To prevent back injuries, they should always use safe lifting techniques and use hoists if necessary. Material should be stacked properly and secured.

Housekeeping Hazards and Controls

One of the simplest methods available to reduce the frequency of unanticipated hazards on a construction site is to enforce a clean and orderly approach to work. Third-party auditors such as workers' compensation and liability insurance representatives will take special notice of a sloppy work environment because they know from experience that there is a high degree of correlation between a sloppy workplace and higher rates of injuries and illnesses.

A frequently cited violation in construction is a lack of housekeeping. Therefore, project management should ask the following:

- Are materials stored in a manner in which they could tip over and hurt someone?
- Are materials or equipment placed so that mobile and mechanized devices have to drive around and perhaps travel closer to employees on foot than needs to be the case? If that is happening, are the affected employees being notified that they need to be on extra alert for mobile devices crossing their paths?
- Is flammable trash such as shop towels soaked in gasoline being tossed into a scrap pile that will be more likely to ignite because of the inclusion of flammables?
- Are flammables being stored nearby operations involving ignition sources, such as welding?

If the answer to any of these questions is yes, then housekeeping efforts should be addressed as soon as possible.

Materials and equipment should be stored so that they will not become a hazard. Enough anticipated hazards exist to deal with on a construction site; creating additional hazards makes the goal of safety more difficult to achieve.

Another reason to keep things orderly because of what may be underneath a pile of materials. Unrecognized hazards may present themselves once the materials are moved or manipulated. There may be an existing hole in the deck that, while too small for a person to fall through, could still cause a trip hazard and potential ankle or knee injury or worse. Once these materials are removed, a small floor hole is easy to overlook. Any holes in walking surfaces should be dealt with prior to beginning work or immediately when they appear as a result of ongoing work activities.

Floor hole covers are a unique form of fall protection. They are not considered fall restraint or fall arrest and do not require advanced competency to successfully employ. Many floor hole covers are used for small holes that a person cannot fall through but still present a trip hazard. For larger floor holes that a person can fall through and cannot be covered over due to the work involved, a more advanced fall protection technique such as guardrails or a PFAS must be employed.

When using a cover is acceptable, a few guidelines must be followed. The cover must be secured so that it remains in place. If mobile or mechanized equipment will be traveling over it, then the cover must be strong enough to support the load; OSHA requires at least twice the maximum axle load of the largest vehicle expected to travel over it. Covers should be readily visible because a cover will sit up as much as an inch or more above the working surface, and an employee could catch a toe on it. Covers should be marked as "HOLE" or "COVER" or painted a specific color to indicate their purpose (ideally with a high-visibility color such as orange) that will be easy for employees to see from a distance.

Whichever method is employed to deal with a floor hole, employees should always be made aware of the methods selected and be prepared to promptly and reliably eliminate the hazard.

Hazards and Controls for Powder Actuated Tools

Powder-actuated tools operate like loaded guns; therefore, they are also called explosive actuated fastening tools. When the trigger of a nail gun is pulled, the firing pin strikes a propellant chemical charge, an explosive charge at the base of a blank firearm cartridge. The charge ignites, causing gases to be rapidly released and build pressure that accelerates the nail through the muzzle and into the substrate surface. This firing mechanism can be either high or low velocity. However, powder-actuated tools with a velocity higher than 328 feet per second (328 ft./s) are not made or sold in the United States.

Nail guns are powder-actuated tools used in construction to fasten materials to hard substrates like steel or concrete. They are made of heat-treated steel, and they use special nails also made of special steel. Some common types of powder-actuated tools are the Hilti and the Ramset gun.

Although powder-actuated tools have built-in safety mechanisms that prevent them from being fired unless the muzzle is pressed firmly against the firing surface, the tool could slip and strike the operator or nearby workers. Powder-filled cartridges could explode, causing the fasteners to become projectiles, or cartridges could ignite flammable gas in the air or liquid in the work area.

If a powder-actuated tool is fired into material that is hard or brittle, the material may chip or splinter, causing it to become a projectile or causing the fastener to ricochet. Powder-actuated tools can also cause damage to materials or property if the fasteners are too long and pierce all the way through the material.

To prevent injury when working with powder-actuated tools, only workers who have been trained and certified to use them properly and safely should operate them, and workers should always wear personal protective equipment. Workers in the vicinity should also wear protective equipment to protect them from misfires. The tool should not be pointed at people, even if it is unloaded.

The muzzle of the tool should be equipped with a guard around the barrel to prevent any material from being projected from the firing surface. Powder-actuated tools should not be used when explosive or flammable gas is present, and they should be inspected before each use to ensure they are not defective. The cartridge should never be loaded until the tool is ready to be used.

Hazards and Controls for Hand and Power Tools

Hand tools are tools powered by hand. Using tools in poor condition or around flammable substances can be hazardous.

Improper use, such as failing to close the jaws of a wrench, can cause tools to slip and injure someone or cause the heads to break and fly off.

If the wooden handles of tools such as hammers are cracked, splintered, or loose, they can break when used, causing the heads to fly off and injure the operator or nearby workers.

Metal tools can produce sparks, which can ignite flammable liquids or gas, and they should not be used if flammable substances exist in the work area.

Power tools are powered by electricity. These tools present burn and electric shock hazards that can lead to serious injuries.

Tools with mushroomed heads present the risk of shattering when they are used, sending sharp metal fragments flying off.

Mushroomed Head

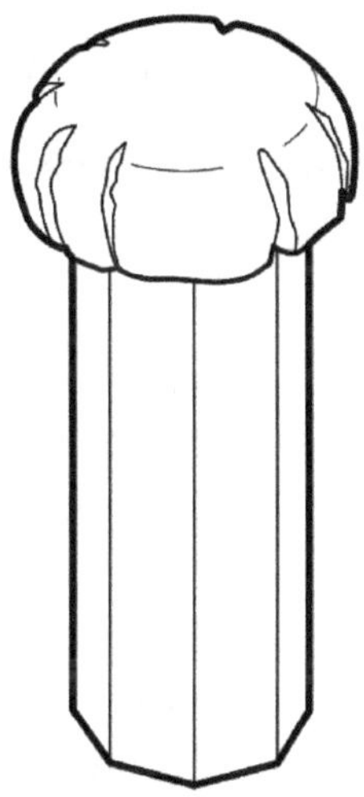

All tools should be inspected before each use to ensure they are in good condition, and workers should wear protective equipment. Power tools should never be used in damp locations, and electrical cords should be disconnected with care and properly maintained to prevent fraying or damage that could cause electrical shocks.

Asbestos Exposure Hazards and Controls

Asbestos is a mineral used in different products primarily for its heat-resistant traits:

In its five forms, asbestos covers the spectrum of toxicity, flammability, irritant, and carcinogen:

- **Chrysotile**: The most abundant type of asbestos found in U.S. construction products; ubiquitous in floor- and wall-related materials and is white or gray in color.
- **Anthophyllite**: Used for its acid-resistant traits.
- **Amosite**: Brown/tan colored; applications are in insulation and exterior surface coating.

- **Crocidolite**: Blue in color, and applications are in filtration systems and insulation.
- **Actinolite and Tremolite**: Contained in mineralogical deposits such as vermiculite, which is used in insulation materials.

There are different types of asbestos, and all are thought to be the cause of certain short-term and chronic illnesses such as asbestosis and mesothelioma. Asbestos particles inhaled by a worker may lead to illnesses such as cancer. The CHST must know the exposure limits for asbestos, have a plan in place, and closely monitor progression of work, specifically in older building renovations, new builds colocated with older ones, and demolition projects.

Exposure limits for asbestos are as follows:

- Not to exceed 0.1 fibers per cubic centimeter (f/cc) of air during an eight-hour work shift
- Short-term exposure not to exceed more than 1 f/cc over half an hour
- Unless a negative exposure assessment at the worksite verifies a below-permissible exposure limit (PEL) environment, a daily protocol for testing with the requisite documentation is required.

Exposures to asbestos fibers have been linked to an array of lung diseases, including cancer. Owners of facilities must be aware of areas containing asbestos. Those performing work in the hazardous areas where asbestos exists must be informed prior to beginning work.

Lead Exposure Hazards and Controls

Lead is an elemental metal that is formed into numerous compounds. It is a dangerous substance and has caused, and continues to cause, both short-term and chronic illnesses in humans. Even minimal exposure can have a chronic and negative impact on a worker's health and capacity for employment.

Lead is known to affect both male and female reproductive systems with the potential for stillbirth or miscarriage for the mother. Surviving offspring have an increased chance of mental or behavioral disorders.

Lead enters the system by ingestion or inhalation, and a significant portion of this will invariably end up in the circulatory system. Lead is typically utilized in plumbing, paint, liner application, and conduit materials on the construction site; workers will find the highest potential for lead exposure during the renovation or demolition of older buildings. The construction industry lead standards include metallic lead, organic lead soap, and organic lead compounds.

Maximum PEL for worker lead exposure is fifty micrograms of lead per cubic meter of air during an eight-hour period. Exposure must be proportionally reduced for any time spent that exceeds the eight-hour period. Established lead standards on the construction site apply to any work activities involving potential lead exposure to a worker; some of these activities may include material installation, structural salvage or demolition, renovation, materials removal, and transport and storage. Employers are responsible for a lead exposure assessment prior to project commencement.

Utilized in paint and as an additive for steel, lead is an inhalation and ingestion hazard whether manifested as fumes (when it is volatilized) or dust (created by abrasive action). The OSHA-**permissible**

exposure limit (PEL) for lead is 50 ug/m^3 during an eight-hour **time-weighted average** (TWA). The OSHA **Action Level** (AL) is 30 ug/m^3 per eight-hour TWA.

Noise Exposure Hazards and Controls

Noise in the worksite can affect the worker on a short- or long-term basis; although noise-related damage may be prevented, it cannot be repaired once it has occurred. Hearing damage interferes with the reception of high-frequency sounds and will be detrimental to the worker's ability to communicate in and out of the worksite. As a result of the known effects of a high-noise environment, measures have been taken to mitigate associated hearing damage. To this purpose, the following maximum exposure limits apply: It is recommended the human ear should not be subject to decibel counts exceeding ninety dBA (OSHA) over an eight-hour period. Shorter exposure times at higher decibel counts can cause instant or long-term damage.

To reduce hearing hazards in the workplace, a noise hazard control process can be implemented with the following steps:

- Noise reduction through use of quieter equipment.
- If feasible, move and arrange worksite equipment away from workers so they may be isolated from noise-producing machinery or tools.
- Engineering controls may be used to provide sound barriers to isolate workers and reduce noise hazards.
- Hearing protection should be worn.

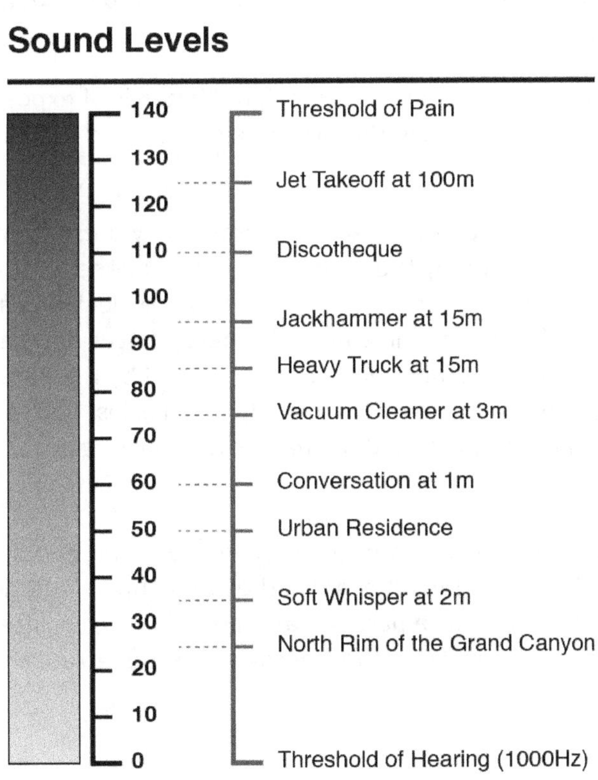

Noise is also considered a stress-inducing agent to be controlled with prescribed methods. OSHA mandates a 90 dBA limit during an eight-hour workday, although NIOSH recommends a lower limit of 85 dBA for the same amount of time. NIOSH does not decide the regulations; instead, they carry out specific workplace studies and make the workplace recommendations resultant from these studies to OSHA. The CHST must be aware of worksite noise levels and must know the procedures and means to reduce worksite levels to ALARP. Known control methods will be utilized to reduce the risk and deal with the hazard specific to its location and nature. The CHST, after identifying and assessing the noise hazard, will work with other safety and operational personnel to decide which control measure (machine relocation, ear plugs, noise barrier, HSE posted signage) will reduce the hazard to a manageable level (for decibel level and time of exposure).

The CHST must also know that OSHA states that if the ambient noise increases by 5 dBA or more, the exposure time for a worker must be cut in half; NIOSH recommends doing the same, albeit based on a 3 dBA increase in ambient noise.

Radiation Exposure Hazards and Controls

Radiation is energy moving through space manifested as electromagnetic waves or as subatomic particles. Ionizing and non-ionizing are the two primary forms of radiation.

Ionizing radiation is the product of electrically charged particles (ions) colliding with and reacting to materials. Ionizing radiation occurs naturally (sun, space) but is also found in things humans build and utilize (x-rays, nuclear power plants). X-rays, gamma rays, alpha particles, beta particles, and neutrons are all sourced from ionizing radiation.

Non-ionizing radiation cannot energize atoms, but it can create the movement and friction necessary to create heat. Non-ionizing radiation comes in the form of lasers, radio frequency (**R/F**), ultraviolet (**UV**) lights, and R/F microwave signals.

The CHST should control measures to reduce the associated hazards of exposure, focus to reduce the time of exposure, increase the distance from the radiation source, and implement appropriate shielding of radiation sources.

Ionizing and non-ionizing are the two types of radiation potentially present on the construction site. Both are forms of energy differentiated by the amount of kinetic energy emitted. Ionizing radiation is manifested in particles or waves with sufficient energy to displace electrons from atoms. In its quest for stabilization, a radioactive atom's nucleus will emit high-energy waves and particles during the process of radioactive decay. Decay can be natural (radon gas is produced by the natural decay of radioisotopes) or man-made (splitting the atom at a nuclear reactor). Alpha particles, beta particles, and gamma rays are the primary radioactive materials produced by the decaying process.

Alpha Particles

Positively charged, high-energy particles comprising two neutrons and two protons. They are the product of decay in heavy radioactive elements, and although highly energetic, they move through the air slowly, as they are heavy in mass. These particles are more dangerous internally (swallowed or entering the bloodstream) than externally, as they will likely not penetrate the skin and will not penetrate through a sheet of paper.

Beta Particles
Like alpha particles, these can move through the air for considerable distances and are naturally occurring or man-made. These fast-moving electrons can, in fact, be stopped by clothing layers or, specifically, thick solid substances such as aluminum, and are most harmful when introduced internally.

Gamma Rays
X-rays are a form of gamma rays and are the primary form of man-made radiation exposure. Gamma rays are clusters of energy of a weightless nature (photons) and are frequently transported with the alpha and beta particles that are emitted from an atom's nucleus. Most gamma rays directed at the human body will simply pass through, but a few of these particles will typically be absorbed. Gamma rays are much more penetrative and will travel through several feet of concrete or several inches of lead before they are stopped.

The average American may be exposed to as much as 360 millirems per year (a **millirem** is one thousandth of a rem unit that measures exposure to radiation) either from external or internal absorption or ingestion. Radiation introduces unspent energy into body tissue, which might damage or kill cells or be causal in the development of cell abnormality.

Acute or chronic exposure to radiation consequently varies, and effects can be seen immediately or after long time periods; damage to the body will depend on radioactive energy absorbed by the body, time of exposure, dosage rate, and the parts of the body that were exposed.

Acute Exposure
Exposure to a single, large dose of radiation or concentrated exposure to a smaller amount in a brief time period.

Chronic Exposure
Exposure to small amounts of radiation over a longer time period.

The control methods for ionizing radiation include shielding, time, and distance from ionizing radioactive sources.

Non-ionizing radiation does not possess the energy to remove electrons (to ionize) from atoms or molecules, hence its difference from ionizing radiation. Ultraviolet (UV) (the sun), lasers, microwaves, and infrared are typical examples of non-ionizing radiation.

Control methods involve measures taken to protect workers from the effects of non-ionizing radiation (sun exposure, welding, heat lamps, and microwaves).

The control hierarchy must be employed by the CHST to assess and assign the control to a specific task where ionizing radiation or non-ionizing radiation presents potential exposure risks. This can be PPE, barriers, distance from source, or any controls deemed appropriate to the task. The CHST will assess the identified hazards for each task. Mindful of the fact that every task is indeed unique, the CHST monitors the development/application of task-specific job safety analysis (JSA) and the ongoing vigilance for associated workplace hazards.

Silica Exposure Hazards and Controls

Silica, like asbestos, can cause serious and long-term lung ailments. Quartz materials are the primary source of crystalline silica and are abundant in a number of construction materials such as mortar, cement, granite, bricks, tile, and sand. Mitigating measures include an assessment of the work environment to determine the proper exposure limiting measures, which may include ventilation, worker position alternatives, tool modifications, controlling worker exposure with time on-site, and respiratory gear. As the construction worksite can be a fluid, shifting environment, it is essential the CHST constantly monitors the often-changing on-site work conditions.

Quartz is the most common form of silica and is contained in sand, clay, granite, gravel, and other types of rock; it is one of the more common minerals on the planet. Exposure to silica can occur in abrasion-related work such as sanding, scraping, grinding, drilling, or cutting. The extent of any detrimental health effects resulting from silica depends strictly on the size of the airborne particles ingested over the period of time the worker is exposed to the material. The most common illness associated with silica is silicosis; like hearing loss, it is preventable but not curable and quite serious, as it creates scar tissue in the lung that inhibits the extraction of oxygen from respiration.

The following are three classifications of silicosis. *Chronic silicosis* is defined by exposure over ten or more years. **Accelerated silicosis** is defined by increased exposure over a shorter time period (five to ten years). **Acute silicosis** is defined by a concentrated exposure manifested in a few weeks or up to five years. Beach sand is an example of very large silica particles that are mostly harmless simply because of their size.

Silica is an irritant, is classified as a carcinogenic, and can actually be flammable given sufficient airborne density and an ignition source. As particle size and density for airborne silicates are produced relative to the task being undertaken, the CHST must assess the circumstances and proper controls with which to mitigate the hazard, while the employer is accountable for providing such resources.

Chemical Exposure Hazards and Controls

Chemical hazards at the workplace are constantly changing as newer chemicals are introduced. Chemical hazards take the form of dust, fumes, gases, vapors, and liquids.

Chemicals take many routes to the body's interior, as they can be ingested, inhaled, or absorbed into and through the skin membrane. The daily habits of workers such as smoking and eating can introduce chemicals into the body when proper hygienic measures are ignored. Inhalation is the most common form of chemical interbody access.

Some chemical agents the CHST must be aware of include ozone, nitrogen oxide, carbon monoxide, chlorinated hydrocarbon solvents, mercury, fluoride, and cadmium.

The CHST must consult SDS in order to be aware of the properties of the chemicals in use. Consultation with the SDS for chemicals to be used will provide information for the material including physical properties (flammability, vapor pressure, and molecular weight), reactivity, toxicity, irritability, and flammability.

The CHST should monitor work activities to ensure chemicals are used and stored properly; all chemicals to be used for a specific task will be listed on the job safety analysis (JSA) with the SDS information included to determine proper usage and application for each specific task.

All chemicals will have applicable PEL, **recommended exposure limits** (REL), and **time-weighted average** (TWA) concentration. The CHST will use available resources and information to determine the proper usage and application for each chemical.

Fumes

Solid particles formed during the vaporization of a metal that cool to room temperature. The fumes produced by welding are an example of this. Airborne concentration of fumes is measured in **milligrams per cubic meter** (mg/m^3), or in the case of lead, the measurement is in **micrograms per cubic meter** ($\mu g/m^3$).

Vapors

Gases that are typically suspended in a liquid state when at room temperature and pressure but go through an evaporation process; these are quantified in **parts per million** (for perspective, one part per million is roughly equivalent to four drops of liquid in a fifty-five-gallon drum).

Mists

Liquid droplets suspended in air due to their small size; measured in the same units as fumes.

Gases

Material molecules suspended in the air at room temperature; these are measured in parts per million in air concentration.

Dust/fibers

As with the silica particles discussed earlier, dust and/or fibers are created by abrasive or percussive work activities such as demolition, drilling, cutting, and sanding. Dust is measured by its airborne concentration in the same units as fumes and mist, although fibers are measured in **fiber per cubic centimeter** (f/cc).

The route of entry to the body for chemicals is through ingestion, such as swallowing as a result of hand-to-mouth actions (eating or smoking in contaminated area), absorption (penetration through the skin), inhalation (primary route of entry for chemicals), and injection (entry through a skin puncture). Dust particles that are inhaled are generally ten microns or less (the diameter of a human hair is approximately 110 microns). Where dust particles are known to be ten microns or less, a respirator with **high-efficiency particulate air** (HEPA) filtration is recommended.

Extreme Temperatures Hazards and Controls

Construction workers can be exposed to extreme cold or extreme heat when working outdoors or in indoor sites without climate control. Working in extreme temperatures presents many hazards to workers' health such as tissue damage, frostbite, heat exhaustion, and heat stroke.

Extreme Cold

Working in extreme cold exposes workers to the risk of tissue damage and cold-induced illnesses such as hypothermia and frostbite. When the human body is exposed to extreme cold, it becomes stressed, and blood is rushed from the extremities to keep the person's core temperature warm. Over prolonged periods of time or in extremely cold temperatures, this can cause frostbite, which can lead to permanent tissue damage. If body temperature drops below 95 degrees, hypothermia can occur. Severe hypothermia can cause shivering, dizziness, loss of consciousness, and, if not treated in time, death.

To prevent frostbite and hypothermia, workers should always keep their heads and hands covered and wear insulating socks and clothing to prevent heat loss. In cold temperatures accompanied by high winds, they should protect their eyes and faces. Workers should drink plenty of water to prevent dehydration.

Extreme Heat

Workers exposed to extreme heat are at risk of heat-related illness such as heat rash, cramps, exhaustion, or heat stroke. Exposure to direct sunlight can cause sunburn, which can be severe enough to cause blistering and permanent tissue damage. Heat rash occurs when the skin becomes inflamed because sweat glands are blocked.

In addition to skin damage, if the body continues to be exposed and lose fluid and electrolytes, heat exhaustion can set in and cause muscle cramps, sweating, rapid pulse, headache, and vomiting.

To prevent heat-related illness, workers should wear lightweight, sweat-wicking clothing and hats to protect them from sun exposure and sunburn. Wearing cooling vests can also help eliminate overheating. They should drink water regularly to prevent dehydration.

Vibration and Impact Exposures Hazards and Controls

Many power and pneumatic tools can produce hazardous vibrations. These tools include air and electric hammers, hammer drills, tampers, chisels, chain saws, and grinders. Long-term use and exposure can result in **Hand-Arm Vibration Syndrome** (HAVS). Symptoms of HAVS include circulation disruption, hand and finger numbness with loss of dexterity, muscle and bone disturbances, and compromised performance in detailed tasks. The **National Institute for Occupational Safety and Health** *(NIOSH)* has threshold limits for hazardous vibration. Limit parameters are based on time and acceleration (duration of exposure and directional frequency of vibration). A control measure for hazardous vibration is to provide anti-vibration gloves as PPE. Factors that limit their effectiveness are glove type, hand size, and grip strength among different workers.

Globally Harmonized System of Classification

The **Globally Harmonized System** of Classification and Labeling of Chemicals (GHS) unifies the way countries define, classify, and communicate safety information about chemical products in the workplace. GHS always uses the same format for labels and safety data sheets. The purpose of GHS is to globally regulate hazardous chemicals and to facilitate trade, as well as to increase efficiency and reduce costs. GHS consists of two components: classification and labeling.

GHS classifies chemicals by three types of hazards: physical, health, and environmental. Chemicals in each class are assigned a category and a number or letter that corresponds to the severity of the hazard.

Health hazards pose a risk to health and include toxins, carcinogens, irritants, or corrosives that cause damage to tissue, respiratory system, or internal organs.

Physical hazards pose a risk to the physical body or the work space. These include explosive or flammable liquids, solids, or gases, oxidizing and pyrophoric chemicals, and self-reactive substances.

Environmental hazards are chemicals that can cause damage to the environment, such as to aquatic or plant life, or the ozone layer.

GHS requires that certain information be included on the labels of all chemicals in the workplace. This information includes the class and category, hazard and precautionary statements, and a signal word.

The hazard statement is a standardized statement that describes the chemical hazard based on the class and category of the chemical.

Like the hazard statement, the precautionary statement is standardized and describes the steps to take to prevent or minimize exposure to the chemical from improper handling.

Chemical labels include signal words such as **Danger**, **Warning**, or **Caution** to communicate the level of hazard of the chemical, based on the chemical's class and category. These signal words also appear on the Safety Data Sheet.

In addition to signal words, GHS labels include pictograms for each chemical. Framed in a red border, the pictograms visually represent the type of hazard the chemical presents.

The **Safety Data Sheet (SDS)** in GHS serves the same function as the now obsolete Material Safety Data Sheets (MSDS) for chemicals in the workplace. Effective December 1, 2013, the MSDS was replaced by the SDS in the United States. These data sheets include sixteen sections that contain standardized information about each chemical, including the class, category, and signal word, as well as more detailed information about the hazards and precautions for safe handling and storage.

Basic Safety Through Design

Many hazards are introduced in the process-planning stages of a project. Without being addressed, these hazards are further magnified as different processes come together on a job site. Addressing safety issues in the active work stage or post-design involves making changes to address hazards that are *now inherent* to a system. Actions at this stage are far less feasible and practical to implement and add costs beyond those of the original design. Post-design modifications must also consider deficiencies in redesign that cause controls to be error-provocative as well as workers' abilities to defeat them.

The practice of safety through design involves integrating hazard elimination and mitigation techniques into the early design stages of a project. This requires both collaboration and cooperation among host employers, contractors, and subcontractors. The process must encompass building layout and design, equipment, materials, tools, energy sources, and work procedures, among many other elements. From a results standpoint, these efforts can be rendered largely ineffective if they are not included in contract documents. This also applies to the written safety programs, policies, and associated materials that define and outline how the respective participants practice safety.

Safety processes should be integrated during the initial phases (conceptual and design) of a project with the intent of binding parties to a desired level of safety performance. Contract parties are far more likely to practice vigilance in safety when responsibilities are assigned and ratified through signed agreements. The level of safety performance over a period of time is directly proportional to proactive implementation from a project's inception. Organizations and site safety practitioners can achieve success in safety through design by applying the following general principles:

- Make safety a priority through management commitment and early integration into design processes.
- Establish hazard/risk assessment methods early, in conjunction with the proper application of the hierarchy of controls.
- Ensure that site safety plans and procedures are part of contract documents.
- Conduct updates and meetings to maintain safety/work alignment throughout the design stage.

The major benefits of improving safety through design include eliminating and significantly reducing worker injury, illness, and environmental damage. Lower rates of incidents amount to improved safety performance, which, in turn, results in lower medical costs and insurance premiums. Other positives include better quality, improved productivity, and avoiding expensive retrofits due to deficiencies in the original design.

Risks with Multiple Trades Working Simultaneously

Construction sites often involve multiple work processes that create hazards through the clashing with or interference from other work activities. This can negatively impact safety and health, equipment and material use, and compatibility, or cause scheduling conflicts. Any of these can result in injury and illness, asset loss, or damage to the public or the environment. Such hazards can be eliminated or minimized through communication, coordination, and strategic planning with monitoring and modification of schedules.

The first step to minimizing hazards of simultaneous operations is to identify all the processes that have the potential to contribute to hazards and risk. Incidents stemming from simultaneous operations can occur as a result of larger projects and companies with multiple work disciplines and among multiple entities such as the owner, contractors, subcontractors, and vendors. It is critical that these elements are identified in preplanning and design efforts. Examples include the following:

- Fire and explosion resulting from reactive processes such as welding or cutting in the presence of vapors and combustibles
- Schedule conflicts where different activities occur in the same place at the same time
- Interferences resulting in delays and failure to meet/achieve contract deadlines
- Events where clash and interference could compromise worker safety and project success

Organizations should identify members within a workforce and among multi-employer sites to represent their respective organizations and assign them roles and responsibilities specific to addressing risk associated with simultaneous operations. Design plans should include meetings to identify hazards and risk and to discuss the controls that would mitigate these risks. To this end, each party should develop designs and documents that cover the following:

- List of clearly defined roles and responsibilities
- List of work processes, procedures, and materials and equipment used
- Summary of all hazards identified for each operation
- Summary of risk assessment techniques and assessment results using established methods that apply a hierarchy of controls and include a summary of proposed controls
- Description of general site conditions and change management procedures to monitor changes to the site conditions and modify processes
- Description of the methods of communication between entities, including equipment to use and contingency plans in the event that communication is lost
- List of loss control techniques and incorporated security measures
- Review of emergency action plans and routes of evacuation where applicable

A document or individual documents should be developed for the purpose of interfacing and prioritizing the elements among parties involved in simultaneous operations. Interface documents should be included as part of multi-employer meetings and reviews.

Principles of Ergonomics

Construction activities present many ergonomic hazards that can result in crippling injuries. These hazards stem from: using hand and power tools; processes that require static postures and positioning; improper body mechanics while pushing, pulling, or lifting; repetitive stress; and elements between workers and material that cause slippage as a result of friction and grip loss.

Construction activities often present a vast array of ergonomic hazards. Lifting and carrying of heavy weight and unbalanced loads result in sprains, strains, and back injuries. Pressure to meet schedule deadlines adds physical and mental stress for workers. Ergonomic injuries and illnesses stem from factors such as poor process design and inadequate training in lifting techniques for proper body mechanics. Repetitive motion, friction coefficient between materials (worker/machine/material interface), and static positions including squatting and kneeling for floor work are major contributors of musculoskeletal disorders and various injuries and illnesses.

For addressing hazards, the underlying **principle of ergonomics** is to fit the needs job to the abilities of the worker. OSHA currently has no ergonomic standards regarding construction and CFR 1926 but, as with many situations, concerns can be cited through *OSHA's* **Section 5(a) (1) General Duty Clause** which states, "employers are required to provide their employees with a place of employment that is free from recognizable hazards that are causing or likely to cause death or serious harm to employees."

Ergonomic hazards can be addressed through the development of an effective program. Such a program should be geared toward ergonomics and contain the following core elements:

- Management commitment with clear objectives and performance reviews
- Employee involvement with communication and feedback
- Detection of hazards through JSAs
- Establishment of hierarchical controls
- Medical management related to ergonomic issues
- Training, monitoring, and program enforcement

Conducting JSAs is vital for applying ergonomic principles to construction processes. The presence and variety of changing elements require that job analyses be performed in conjunction with other procedures that acquire more information than standard JSAs. Elements should include the following:

- Observing the worker performing the task
- Taking photos and videotaping
- Conducting interviews with workers
- Conducting surveys and questionnaires
- Making calculations of force requirements, reach distance, and material positioning

Once the initial hazard survey is complete, the CHST will have the information necessary to determine the most effective measures including equipment adaptations, alternate material handling practices, and their worksite modifications. It is necessary for the CHST to be present in the worksite in order to mentor workers and ensure the agreed-to ergonomic procedures are adhered to. All worksite applications and procedural changes must be documented, disseminated, and properly communicated to all levels of personnel with all available methods. When an ergonomic problem or hazard is encountered on the worksite, its identification is followed by assessment and mitigation. Solutions can be implemented according to the priority of the issue. Immediate hazards that do not constitute major process changes should be made immediately. More difficult hazards and problems that involve process changes should be associated with a CAPA plan.

When an ergonomic problem or hazard is encountered on the worksite, its identification is followed by assessment and mitigation. Assessment provides the information needed to prioritize the hazard and allocate the resources deemed relative to its risk factor (on a standard risk matrix, this will be low, medium, or high). Mitigation of this hazard will assign the control appropriate (elimination, substitution, engineering, administrative, or PPE) to close the hazard out. Again, whether or not a formal CAPA plan is required will need to be decided by all project management personnel and the CHST.

Personal Protective Equipment

In all cases, the employer must assess the workplace for hazards and establish controls that reduce the requirements for **personal protective equipment** (PPE). The employer should select and provide the appropriate PPE, determine when it's to be used, and provide employees with training for proper use. Before PPE is considered, all practical methods of substitution, engineering, and administrative controls must be explored. Although PPE can modify an incident's degree of consequence, it must be noted that it does not reduce hazard risk and, therefore, is not a control.

Policies and programs must have consistent monitoring and enforcement to ensure their effectiveness. Employees are responsible for using PPE in accordance with the training provided and for maintaining its proper condition through daily inspection and cleaning.

Following is a list of the major types of PPE, their function, and the applicable criteria for construction activities:

- **Hard Hats**: Head protection is used to protect the worker from trauma, impact, and voltage hazards. ANSI criteria are Type I (for top impact and spinal protection) and Type II (for side impact protection in addition to having Type I criteria). ANSI-rated hard hats contain relevant information including year, type, and class. ANSI adopts the manufacturer's recommendations for component replacement and expiration.

- **Safety Glasses and Goggles**: Eye protection is used to protect flying particles, dust, splashes, and infectious substances. The tinting of laser safety goggles protects eyes against the intense light of lasers.

- **Face Shields and Welding Shields**: These protect the face from dust, sparks, and hazardous sprays. Welding shields protect the face from sparks and the light emissions from welding, cutting, and brazing operations. The degree of shading/tinting depends on the weld type.

- **Ear Plugs and Ear Muffs**: These are for noise attenuation and to protect hearing from occupational noise exposures. Compliance is to an 8-hour, time-weighted average at 90 dBA (A-weighted decibels) for hearing protection, and 85 dBA for the requirement of OSHA's Hearing Conservation Program. Protection devices provide a noise reduction rating (NRR) number, which is applied to various formulas and computed to achieve noise attenuation to compliant levels of exposure.

- **Respirators and Associated Equipment**: Various respirators include filtering face pieces, half-masks, self-contained breathing apparatus, and supplied air systems. Respirators are used in conjunction with filtering cartridges that are color-coded for the appropriate level of protection and conditions. Caution should be taken in situations of voluntary employee self-protection with dust masks and filters. This action is a trigger point for the employer to develop a written program.

- **Gloves**: These are used to protect the hands against abrasions, cuts, fractures, amputations, burns, and chemical exposures. Gloves can be made of canvas, leather, metal meshing, coated fabric, DuPont Kevlar® fiber, and rubber. Rubber gloves of nitrile and butyl are effective against chemicals and permeation respectively, while Kevlar® is best for cutting operations.

- **Boots**: These are for protection against heavy, falling, or rolling objects. They are also used for protecting against sharp and piercing materials, wet and slippery conditions, and heat and molten materials. OSHA regulations require that safety toe footwear meets the standards and criteria set by ANSI.

- **Body Protection**: This is for protection against hot environments, cuts, impacts, intense heat sources, radiation, or as part of respiratory protection. Body protection includes vests, jackets, coveralls, and full-body suits of prioritized hazard classes A, B, C, and D (highest to lowest) for type and level of protection.

Basic Testing and Monitoring Equipment

Monitoring equipment can surveil and indicate dangers before work is commenced in confined spaces and other worksite areas requiring prior-to-entry testing.

Equipment maintenance and calibration is part of equipment monitoring. Field calibration is performed to ensure monitoring equipment is functioning properly. If field calibration is not successful, the manufacturer may be consulted. It is important to note that most units have a mandated annual inspection and calibration; **certified temperature calibration** (CTC) is a resource providing calibration and annual certification of monitoring instruments. Any questions or concerns about a monitoring unit can be directed to the CTC website.

Real-time instruments or monitors are available (direct reading monitoring instruments). Experienced personnel use these to detect an oxygen-deprived or oxygen-rich atmosphere and airborne contaminants or flammable atmospheres to assess the safety of a confined area in real time.

Electrical testing monitors inform workers whether electrical lines or power sources are energized and the corresponding amperage and voltage readings unique to each source. There are also gas monitors designed and utilized for single-, double-, or multi-gas monitoring. Multi-gas monitors incorporate separate sensors to measure oxygen, atmosphere combustibility, and if equipped for such, the concentrations of upwards of six gases in a single unit. These units measure aforementioned gases and are equipped with alarms to indicate low oxygen levels, toxic gas concentrations, or monitor malfunction.

Decibel meters assess the level of noise while vibration meters sample the vibration caused by machines and other equipment in use in the work space.

Radiation detectors are used to measure the amount of electromagnetic (non-ionizing) radiation. They are also used to measure ionizing radiation, such as harmful far UV light, gamma rays, and radioactive decay.

Four gas meters sample the air in confined spaces to detect the levels of carbon monoxide, hydrogen sulfide, combustible gases, and oxygen in the atmosphere. The first three gases are monitored for amounts that are too high; oxygen is monitored for oxygen deficiencies.

Industrial hygiene is the study of stresses in the workplace that can cause injury or illness. These stresses include poor air quality, hazardous chemicals, waste, physical agents, repetitive stress injuries, bloodborne pathogens, radiation, and noise pollution.

To identify these stresses, industrial hygienists perform workplace walkthroughs, during which they observe physical characteristics of the work space, interview workers, and consult SDS, inspection reports, and production data to discover which work spaces or processes present the most exposure risk. In addition to physical inspections, to identify hazards from poor industrial hygiene, they use basic testing and monitoring equipment including electronic measuring devices to sample air quality, noise, radiation, gases, and other contaminants.

Hierarchy of Controls

Hierarchy of controls is a long-established protocol for assessing hazards and prioritizing workplace mitigation measures for such. It is necessary to reconcile the mitigation of worksite hazards with the resources that are available and/or allotted to this end. Practical considerations dictating every hazard cannot be completely removed from a worksite, and therefore consideration must be given to prioritizing the hazards with the highest potential harm to workers. The hierarchy of controls is used in concert with a **cost-benefit analysis** (CBA) of worksite hazards to determine resource allocation where they are perceived to be needed the most. The hierarchy of controls is one of the primary tools utilized in the NIOSH initiative for **prevention through design** (PtD). Based on the concept of designing hazards out of the workplace, PtD applies preemptive design changes to work methods, tools, equipment, operations, processes, worker facilities, newer technologies, and workplace organization. The hierarchy of controls includes the following:

- **Elimination**: This is by far the most preferable in an ideal environment and typically the least flexible and most expensive of controls. The optimum time for the application of elimination controls is during the project planning stage while it is still feasible to make major changes. Once work is under way, it is more expeditious to default to administrative or PPE controls than to implement changes using the elimination control. Elimination should be considered a preventive action in the CAPA process.

- **Substitution**: Substitution, while one of the preferred hazard mitigation methods, typifies the more desirable but higher-cost hazard solution. It becomes increasingly difficult and less cost-effective to implement substitution controls once the worksite has already been established. When the contractor chooses the means and controls for a construction project, worksite equipment, materials, and tools have usually been ordered and delivered. Changing any of these resources during an already established project or altering the worksite can be prohibitively expensive. However, if an identified hazard exists, or an incident has occurred that has resulted in financial or human loss, substitution may be the only solution in a corrective action.

- **Engineering**: This covers the middle ground between elimination/substitution and administrative/PPE. It is a method whose purpose is to remove a worksite hazard at its source prior to worker exposure and by design. It is also adaptable in an already-existing construction site. An engineering control generally consists of modifications or barriers designed into a worksite that preclude workers' exposure to hazard. Removing a source worksite hazard is, in respect to the CAPA process, a preventive action. Identifying, mitigating, and eliminating a hazard before it becomes an incident is a proactive, preventive measure.

- **Administrative**: These controls tend to be less expensive than other mitigation measures but are significantly more reliant on worker participation, as they are implemented in areas with decreased hazard controls in place. HSE signage, postings, and training are easily and inexpensively implemented, although more challenging to incorporate. Administrative controls must be part of the CAPA process, as they are required to guide personnel assigned to corrective and preventive tasks in order to effectively implement the CAPA plan on file.

- **Personal protective equipment**: As with administrative measures, PPE is used in areas where hazards are not very effectively controlled. Heavily dependent on monitoring, mentoring, and worker participation, this program is contingent on effective and consistent worksite monitoring by personnel such as the CHST. Standard PPE includes boots, a hard hat, gloves, and safety glasses, but can include other equipment specific to the task at hand. As part of the CAPA process, PPE can be considered both preventive (if used proactively before an incident occurs) and corrective (if used in correction after a hazard has been identified or post-incident).

Using Hierarchy of Controls to Protect Workers

Applying the hierarchy of controls to protect workers requires a specific logic and behavioral flow. This can be conceptualized via:

- **Higher-level controls**, which focus primarily on design, equipment, and materials *at the source.*
- **Mid-level controls**, which seek to reduce existing hazard risk through system redesign and process modifications.
- **Low-level controls**, which protect workers via modifying behaviors and reducing hazard impact *at the point of the worker.*

The most effective means of eliminating or reducing risk requires the understanding that all higher-level controls should be analyzed and exhausted prior to considering lower-level controls because the highest-level controls more greatly affect hazard presence and risk, while the lowest levels affect the consequence of existing hazards. The following includes a prioritized list of hierarchical controls to protect workers.

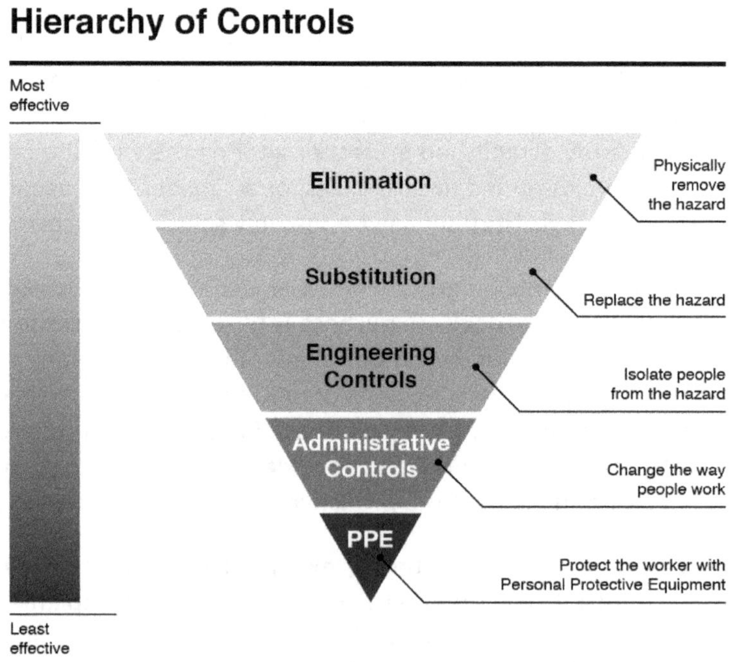

I. Safety Through Design (Aspect of Elimination)

Applying safety through design is the best means of protecting workers as hazard risk is most effectively addressed at the lowest cost. Eliminating hazards here greatly enhances an organization's control of processes, provides worker protection, and reduces risk assessment efforts subsequent to project start-up.

II. Redesign (Aspect of Elimination)

This stage of action requires design and equipment modification to address system hazards. Redesign proposals often present issues with practicality, feasibility, and additional costs. Elements of redesign can include measures such as process automation and extensive fall-prevention systems.

III. Substitution

Substitution involves removing a hazardous process or material and replacing it with a less hazardous alternative. Substitution can be effective for risk reduction, but its reactive nature (post-incident), higher residual risk, and compounding costs make it less effective than design and redesign measures. Examples of substitution include replacing a hazardous chemical with a more inert alternative and introducing a powered conveyor to mitigate risks associated with manual material handling.

IV. Engineering Controls

Engineering controls are components introduced into a system to isolate an existing hazard. This reduces hazard risk by separating workers and processes from the hazardous element. Relative concerns include workers' ability to defeat controls and the ongoing presence of the hazard source. Examples of engineering controls are ventilation systems and machine guarding.

V. Administrative Controls

These controls affect the manner in which workers behave or react to a given situation. They involve avoiding uncontrolled hazard conditions and rely upon worker capabilities and human performance. Examples of administrative controls include selecting specific personnel to perform a task and training workers.

VI. Personal Protective Equipment (PPE)

Although PPE is vital for protecting workers against injury and illness, it is the least effective method of control in the hierarchy. PPE places the element of protection, or barrier, at the point of the worker, can negatively affect performance, and can easily be defeated. PPE comes in an array of forms including face shields, gloves, boots, hearing protection, etc.

Eliminating or Mitigating Hazards

Historically, the construction industry sector has had a highly disproportionate number of workforce injuries and illnesses when compared to other industries. Therefore, policy and plan criteria for construction safety must go beyond compliance as a minimum. Many construction accidents stem from an organization's ineffective structure and the absence of a safety culture. This includes the lack of coordination and communication, as well as deficiencies in detecting existing hazards. These deficiencies result in diminished safety levels, poor performance, and more loss incidents. Hazard awareness and identification are the keys to eliminating or reducing risk. Effective hazard recognition requires training and should be performed by qualified personnel.

A **hazard** is any unsafe act or condition that, if left unchecked, can result in injury and/or illness. Hazards can also negatively affect people, property, equipment, and the environment. Identifying hazards is a vital part of maintaining a safe work environment, and hazard recognition should not fall solely on the shoulders of safety personnel. Hazard recognition, assessment, and mitigation strategies are most effective when there's a strong program and all layers of the workforce are committed, including project or site ownership, management, supervisors, maintenance, and laborers. Hazard assessment and risk reduction efforts can be proactive, in situ, or reactive.

Failure Mode and Effects Analysis (FMEA) is a proactive tool for hazard recognition and mitigation strategies. This process is used in the engineering and design phase to identify and describe potential system failures before an incident occurs. FMEA provides safety professionals with useful information that can help reduce risk in a process. A safety professional compares and contrasts pertinent acquired data with historical data and relevant information to identify and examine potential failures. Mathematical modeling can also analyze reliability and failure rates.

Hazards can be introduced in the process planning stages, and then magnified when different processes come together on a job site. Feasibility, cost, and lack of coordination between entities are major contributors that can introduce hazardous conditions into design processes. Eliminating hazards during a project's design stage is an effective mitigation strategy because it is proactive. It can reduce or eliminate an array of unwanted conditions that would otherwise be overlooked. Built-in hazards have three modes, which are described as follows:

1. **Dormant:** A hazard created by the inclusion or omission of elements. It lies undetected for subsequent arrival.

2. **Armed:** A hazard that's triggered or loaded through the combining or assembling of separate processes.

3. **Active:** A situation where a combination of elements actuates dormant and armed hazards into events.

Hazard identification, assessment, and mitigation is proactively critical to construction compliance. Such review of potential workplace illnesses, injuries, and incidents is required by law. The CHST will assist in worksite efforts to foster an environment of participation-based hazard identification. The success or failure of a project rests on a successful hazard identification program and the documentation that validates it.

Hazards may include, but are not limited to, potential worker threats. First, the workplace must be examined for existing threats. These should be thoroughly documented and mitigated whenever possible. Any and all existing equipment manuals, chemical safety data sheets (SDS), OSHA logs, documentation on previous injuries, worker injury/illness patterns, past safety committee meeting minutes, or any other pertinent documentation must be reviewed prior to project implementation. Neglecting to do so can result in the same hazards, all of which may be avoidable.

After reviewing past hazard documentation, a current inspection of the worksite is essential in identifying and mitigating potential new threats. All equipment maintenance, as well as electrical and chemical hazards, must be assessed. Emergency procedures must be reviewed for accuracy in accordance with the existing workplace. Workplace organization and flow should be examined for current threats as well.

Any and all health hazards must be fully noted and documented. All chemical, electrical, physical, biological, and ergonomic hazards and past incidents should be assessed. Important tools may involve SDS and past worker medical records. Other tools such as air sampling may be required according to the planned project.

Lastly, hazard identification, assessment, and mitigation strategies should include full incident investigations. A process of reporting, documenting, identifying the root cause, and having corrective action plans in place must be part of hazard strategies. It is wise to have both emergency and non-emergency hazard workflows in place identifying personnel responsible, reporting and communication structures, documentation guidelines, incident templates, and corrective/preventive action requirements.

The goal should be to reduce risk and decrease hazards. Some hazard studies indicate risk reduction remains the focal point of project HSE planning, although further studies suggest that decreasing the hazards involved can lower the risk of individual tasks. Hazard identification is a critical tool and is utilized by the CHST to qualify and quantify the most practicable worksite hazard and risk reduction measures.

Hazard assessments must contain clearly identified hazards and mitigation plans for each. Mitigation defines the measures taken and resources allocated toward worksite risk reduction of assessed hazards; each hazard is broken down, analyzed, assessed for its unique risks, and has the proper mitigating measures applied and documented. Failure to conduct hazard identification, assessment, and mitigation will lead to undesirable, unforeseen, and reckless construction.

Following up on Corrective Actions

As part of design safety planning, corrective actions are developed to address hazards in a proactive manner (before work activities begin). The main objective of design safety is hazard prevention through developing controls that eliminate risk. Risks that cannot be completely eliminated must be reduced to a level as low as reasonably practicable. Controls are developed as a result of hazard detection where risk analysis, prioritization, and a hierarchy of controls were guiding factors in decision making and intended outcomes. This means that the corrective action was developed with an expectation of performance and effectiveness in mind.

Corrective actions, whether pre- or post-design, often require performing cost-benefit analyses that consider the input of money and level of effort and theorize the point in time where a return on investment (or monetary balance) may be achieved. The question, "Are/were the results of the action worth the time, money, and effort?" further necessitates evaluating control performance and efficiency. Follow-up efforts must be conducted to evaluate the effectiveness of corrective actions.

Evaluating performance through follow-up efforts measures the effectiveness of corrective actions against expectations, or the action's planned level of performance. Results indicate whether or not the corrective actions sufficiently addressed the target hazard and reveal the amount of residual risk. Corrective actions must be analyzed with great caution prior to implementation to avoid introducing a new or additional hazard into the project design.

Follow-up procedures can be conducted as follows:

- Inspect/review performance of new equipment and materials.
- Assess operational conflicts such as scheduling and interference.
- Review design/layout modifications to reduce ergonomic and fall hazards.
- Incorporate as part of change management (e.g., modified or new equipment, materials, processes, projects).

Following up on corrective actions is vital for effectively managing process safety and risk and continually improving safety for workers, the public, and the environment. Follow-up can result in modifications to worker behaviors and prescribed procedures, improvements to program-level elements, as well as have a positive impact on organizational approaches regarding design safety with future projects. Effective corrective actions and follow-up rely on the combined efforts of employers, the safety professional, design engineers, and other key personnel. Follow-up and other risk management methods should be reviewed and adjusted on a regular basis to ensure continuing efficiency.

Changing Construction Site Conditions

Anticipating workplace changes and potential hazards is a must before project work can begin. Many international, federal, state, and local regulatory agencies demand hazard analysis be performed and documented results kept on file. Most of these records are subject to audits, so it is highly advisable to become familiar with the regulations and statutes with which the project will be held in compliance prior to project implementation.

First, an on-site baseline hazard analysis should be conducted and filed. It is recommended this analysis has been performed within the past 5 years at minimum, but again, construction employers and project management need to be aware of regulations that may differ. A baseline hazard analysis includes the following:

- A review of any and all safety- and health-related records on file and that the present system is reviewed annually at minimum
- A review of any hazard surveillance policy
- A review of any existing engineering controls
- A review of emergency policies and procedures
- A check to ensure that an effective hazard safety reporting system is present and that any communication policies and procedures have been part of employee training
- A check to ensure that change analysis is performed in the event of needful difference in process, facilities, location, weather, equipment, or project materials
- A review of any prior incidents and that root cause analysis (RCA) has been performed

- A review that any corrective and preventive action (CAPA) in response to prior incidents has been documented and implemented
- A check to ensure that material safety data sheets are in place if chemicals are present
- A check to ensure an exposure control plan is in place, if historically applicable
- A check to ensure that all Occupational Safety and Health Administration (OSHA)–mandated programs are in place
- A review to ensure all preventive maintenance is being, and has been, performed as required by policy
- A check to ensure any and all training records are present
- A check to ensure all project documentation includes the appropriate change control (for example, full notation of policy and procedure changes including altered content, date implemented, and version control)
- A check to ensure recommendations for any changes or improvements on the above prior to project implementation are included and, if necessary, implemented and documented prior to project implementation

Construction projects can contain a myriad of necessary changes, whether these are changes performed on the fly or through planned implementation. Changes that may have financial, safety, health, environmental, or scheduling impacts to the project must be effectively analyzed, prioritized, and implemented by project management. As the construction project unfolds, project management is under obligation to perform this change analysis on a regular basis according to established policies, procedures, and regulations. All workplace personnel are responsible for reporting issues via the established reporting system, and change analysis can and should, in part, come from that feedback. All changes that are implemented must be fully documented and kept on file. Neglecting to perform the appropriate hazard and change analyses and keep all documentation related to such may result in citation for noncompliance, significant monetary loss, or suspension of the project.

Job/Hazard Safety Analyses

A **job safety analysis** (JSA) is a written model for pre-task planning that considers all possible hazards related to a particular job where multiple steps must be performed. The process involves reviewing the job steps and methods, detecting associated hazards, and developing corrective actions to eliminate or reduce risk. A JSA format shows the list of job steps down the left-hand column, associated hazards down the middle, and corrective actions down the right column. The tasks included in the JSA should be selected based on priority with regard to hazards experienced or hazard potential. Jobs and tasks should be neither too short nor too long (e.g., five to fifteen steps).

A job safety analysis (JSA) procedure involves reviewing job steps and methods, detecting associated hazards, and developing corrective actions to eliminate or reduce risk. The most logical way to format a JSA is to list the sequence of job steps in the left-hand column, list possible hazards in the middle column, and list corrective actions in the right-hand column. This assures that the three elements and their sub-elements achieve horizontal alignment. Jobs selected for JSA development should reflect a focus on prioritization and level of need with respect to hazard potential. Jobs and tasks should neither be too short or too long (i.e., turning a bolt or wiring a facility). This helps to maintain a sensible number of sequential steps (5 to 15), though there is no established rule.

The following chart emphasizes notable details and formatting requirements associated with the development of effective JSAs:

Itemized Job Safety Analysis (JSA) Instructional Tool		
Job Step Sequence	**Potential Hazards Detected**	**Corrective Actions/Procedures**
1. Job Step (1) *In this column, the job is broken down into a sequence of job steps.* *A job step is one of many parts involved in completing an overall task.* *The job step should represent a portion of a task with a logical beginning and end point.* *In this column, the step should be aptly named (what's being done), and all associated actions and relevant items involved should be listed.* *The next step in the sequence of job steps will be listed in the same manner, in this column, as part of row (2) below.*	**1. Hazards of Job Step 1** *This column represents the hazards associated with each job step. Each hazard, or group of hazards, will correspond with the preceding job step (left).* *When conducting a JSA, hazard analysis efforts should consider the following: elements associated with human actions and movements; perceived hazards from tools and equipment used during the job step; and any possible environmental factors, including nearby processes that are separate from all job elements.* *It is important to consider what hazards are present, what could be introduced, and any associated outcomes related to injury and illness type. The concept of hazard prioritization is helpful for the consideration and inclusion of multiple hazards for each job step.*	**1. Corrective Actions for Hazards** *Recommendations and corrective actions are developed using the first two elements of the row (Job Step (1) and Hazards of Job Step 1). For these measures, techniques of prioritization, feasibility, and a hierarchy of controls should be applied. They can consist of elimination through design and substitution, engineering and administrative controls, and/or the provision of PPE.* *Measures can also be driven by procedures, directions, and directives. Written elements for recommended procedures must be clearly described. Procedures should be suitable for the worker. A corrective action should be given for each hazard identified in a job step, and it should maintain alignment in horizontal formatting with respect to the hazard and the step.*
2. Job Step (2)	**2. Hazards of (Job Step 2)**	**2. Corrective Actions**
3. Job Step	**3. Hazards of Job Step**	**3. Corrective Actions**

The right worker should be chosen to assist in JSA development. The worker should be skilled, experienced, and possess sound communication abilities. A good performer can also serve as a baseline for future monitoring and observations. The following is a list of benefits associated with effective JSAs:

- Continual improvement efforts pre-/post-incident
- Tool for change management (new jobs or alteration of)
- Index for future safety observations
- Hazard elimination/risk mitigation
- Training development, instruction, and retraining

Performing and maintaining JSAs also provides greater knowledge to both supervisors and employees about processes, tools, and equipment/materials required. Information gathered and reviewed before operations can help increase hazard awareness and worker efficiency. Such information can be used in questioning, for incident investigation to determine if the JSA was followed, or if it is deficient. JSAs should be periodically reviewed for effectiveness and possible revisions.

Hazards Based on Level of Risk

The employer is responsible for hazard detection, hazard assessment, and protecting employees from workplace hazards. The process is primarily performed through a basic three-step premise: **Recognize, Evaluate,** and **Control.** This may seem simple, but each step in the process requires a multitude of factors, concepts, and considerations. Control methods should be considered based on those with the greatest effect on the hazards being analyzed. In turn, this produces the most effective impact with respect to relevant outcomes.

A major aspect in establishing a hierarchy of controls is understanding that the result of corrective measures is the achievement of an acceptable risk level. This means that the probability of occurrence, exposure level, and degree of harm are as low as reasonably tolerable and achievable for the situation. Factors including feasibility, cost with respect to expected outcomes, and the potential to create new or additional hazards can impact the means and level for addressing a hazard. Such situations can result in a higher degree of residual risk where reassessment and additional measures may be required. These extenuating elements express a behavioral flow and complications that can arise when addressing hazards and risk. Such factors require a systematic way of thinking when applying a hierarchy of controls.

Throughout the hierarchy, an important aspect to note is that higher controls reflect more effective strategies that are both proactive and preventive, thereby eliminating or greatly reducing risk. These controls rely on safer designs, reengineering equipment, and replacing material elements where corrective actions are directed at hazards inherent in a system. Lower hierarchal controls reflect strategies that are more reactive with respect to incident occurrence. They result in higher expenses that are compounded by incident costs. In contrast to design and engineering measures, these controls are geared toward modifying human behaviors in the presence of hazard risk.

When establishing controls to eliminate or reduce risk, all elements of a higher order should be considered and exhausted before selecting controls from lower in the hierarchy. This presents a more thorough analysis of controls and results in more effective measures.

Practitioners of safety can evaluate hazard risk via risk matrices. Risk matrices function by determining the potential of event frequency (or likelihood of occurrence) versus event severity level to determine hazard risk. This process can be used in conjunction with JSAs and other analysis tools. Risk matrices are derived of alpha (qualitative) or numeric (quantitative) elements to yield an overall risk level or score. Subsequent to determining risk level, appropriate corrective actions are applied to eliminate or reduce risk.

The **Z-10 Risk Assessment Matrix** and other risk matrices from the *American National Standards Institute (ANSI)* provide hazard risk prioritization (for both frequency and severity) from which risk level can be determined. Using this strategy, the rate of incident frequency (or likelihood of occurrence) can be plotted versus the incident severity level to determine the overall level of risk. This and other matrix types can derive **alpha** (qualitative) or **numeric** (quantitative) elements from which a risk level can be attained. Subsequent to determining risk level, an organization should consider mitigation measures to eliminate or reduce the risk. Control measures should be prioritized through a hierarchy and feasibility, beginning with elimination, substitution, and engineering controls. When a hazard is eliminated, the risk is eliminated as well. Efforts that result in reduced risk require additional assessment for the possibility of single or multiple controls. Lower hierarchal controls include warnings, administrative controls, and personal protective equipment (PPE).

Here's a look at two of these matrices:

Severity / Probability	Catastrophic (1)	Critical (2)	Marginal (3)	Negligible (4)
Frequent (A)	High	High	Serious	Medium
Probable (B)	High	High	Serious	Medium
Occasional (C)	High	Serious	Medium	Low
Remote (D)	Serious	Medium	Medium	Low
Improbable (E)	Medium	Medium	Medium	Low

Likelihood / Impact	Rare (1)	Unlikely (2)	Possible (3)	Likely (4)	Almost certain (5)
Catastrophic (5)	5	10	15	20	25
Major (4)	4	8	12	16	20
Moderate (3)	3	6	9	12	15
Minor (2)	2	4	6	8	10
Insignificant (1)	1	2	3	4	5

Emergency Preparedness and Fire Prevention

Fire Protection and Prevention Methods

A CHST should be very familiar with basic fire safety principles, the classes of fires, and the types of commonly used extinguishing agents because many of the materials found in construction are flammable.

Fire extinguishers on construction sites should be inspected at least monthly. This is an NFPA (National Fire Protection Association) consensus standard that OSHA has adopted. The monthly inspection is a simple task that does not require an advanced knowledge of fire extinguishers, and many construction personnel will be sufficiently knowledgeable to be assigned this duty. During the monthly inspection, the inspector should ensure the extinguisher is in an area that is known to all employees in the work area and easy to access. The needle in the pressure gauge should be in the green, which means the extinguisher has appropriate pressure. Any physical damage or other apparent deficiencies that can be identified by visual analysis should be addressed. When in doubt, the extinguisher should be removed from service until it can be more thoroughly inspected by a fire safety professional.

OSHA also requires that fire extinguishers undergo an annual maintenance inspection, which is more involved than the monthly inspection. This is typically performed by a fire safety professional, and any deficient units will be refilled or replaced at this time.

Common extinguishing agents include water, carbon dioxide (CO_2), dry chemical, and various types of foam-like agents. Fire extinguisher technology advances have resulted in refinements to these basic extinguishing agents and the introduction of newer agents, some of which are hybrids of the basic types. When the time comes to conduct the annual maintenance inspection, a CHST may suggest to management the review of newer and more versatile extinguisher types for purchase instead of refilling older units and/or replacing deficient units.

There are five classes of fires, and not all fire extinguishers will be acceptable for use with all of them. A CHST should commit to memory the fire classes; extinguishers will identify the types of fires they are capable of dealing with by a letter designation representing a particular class of fire.

Class A (Ordinary Combustibles) fires include paper, wood, cloth, most trash, and also some plastics and other materials. Many of the combustible materials on construction sites will fall into this category. Water, dry chemical, and foam are suitable for Class A fires. CO_2 extinguishers should not be used for these types of fires because the stream of pressurized carbon CO_2 could displace lighter combustibles, like paper, and inadvertently spread the fire.

Class B (Flammable Liquids and Gases) fires involve flammable liquids common to construction sites such as gasoline and solvents and also some flammable gases like propane. CO_2, dry chemical, and some types of foam are acceptable extinguishing agents, although water is not an ideal agent for these types of fires. That is because some flammable liquids, when mixed with water, will still be flammable even though diluted. Additionally, the stream of water or foam may displace the flammable liquids and spread the fire. However, there are some types of foam-based extinguishers that employ an agent particularly suited to flammable liquids.

Class C (Electrical Equipment) fires should never be fought with any wet extinguishing agents for obvious reasons. CO_2 and dry chemical extinguishers are commonly employed for electrical fires.

Dry chemical extinguishers are common to many workplaces because of their versatility, and a CHST can expect to see them in construction. Many are called ABC extinguishers because of their suitability with ordinary combustible, flammable liquids, and electrical fires; these letters will appear on the extinguisher. **Multipurpose dry chemical** is a phrase that refers to modern dry chemical extinguishing agents. On a modern construction site, **dry chemical** and **multipurpose dry chemical** will refer to the same thing; much older dry chemical extinguishers were suitable for Class B and C fires only and have fallen out of use.

Class D (Combustible Metals) fires involve the alkali family of metals found in the periodic table, such as lithium, sodium, and potassium. These elements react violently with water and ignite when exposed to air. Fortunately, encountering them on a construction site would be highly unusual. Only special-purpose "dry powder" extinguishing agents can be used on these fires, which should not be confused with dry chemical.

Class K (Oil and Fats) fires are those involving cooking oils, animal and vegetable fats, and other materials typically found in kitchen environments, so a CHST will not encounter these in construction.

Fire Extinguisher Classes					
Class	**Material**	**Water**	**CO_2**	**Dry Chemical**	**Foam**
A	Ordinary combustibles	YES	NO	YES	YES
B	Flammable liquids and gases	NO	YES	YES	SOME
C	Electrical equipment	NO	YES	YES	NO
D	Combustible metals	Dry powder only (<u>not</u> the same as dry chemical)			
K	Oils and fats	Wet chemical			

The fire class letter assignments represent simple groupings and should not be confused as representing a greater or lesser hazard; Class A fires are not necessarily more severe than Class B, and vice versa.

When inside buildings during construction activity, there must be no more than 100 feet of travel distance to the nearest extinguisher. In multistory buildings, there must be at least one on each floor, and of those, at least one must be adjacent to a stairway. If the building has a flammable storage room, then a fire extinguisher must be located within 10 feet of the door or entry to the storage room.

For outdoor construction activity, an extinguisher must be located within 100 feet of open yard storage and within 50 feet of five or more gallons of flammable materials. For designated outdoor flammable storage areas, an extinguisher must be located at least 25 feet away, but no farther than 75 feet.

Emergency Action Plans

Each industry type, line of business, type of work, and material used presents different hazards to workers and the environment. Emergency action plans and medical/first-aid procedures must be developed and aligned to meet the needs of each different type of hazard. As a start, emergency and medical/first-aid plans and associated components must meet or exceed the standards outlined in OSHA 1926.34 and 1926.35 for means of evacuation and emergency action and in OSHA 1926.50 for medical and first-aid procedures.

The organization should designate an emergency response coordinator. This individual is responsible for conducting preplanning tasks prior to the beginning of work activities and communicating and coordinating procedural elements between site/facility personnel and local emergency service providers.

Critical steps for effective emergency action plans include the following:

- Review all emergency response procedures and contingency plans with regard to modification of work processes and wherever applicable.
- Determine the process requirements and availability of both on-site and off-site communication equipment.
- Confirm, update, and maintain information relevant to emergency contacts, field and facility evacuation routes, assembly areas, and routes to nearby medical facilities.
- Locate fire extinguisher and exit signs above their respective extinguishers and exit doors and maintain proper housekeeping to ensure clear access to both elements in permanent and temporary facilities.
- Review site conditions, changes to site conditions, operations, and personnel availability for emergency procedural risks and compatibility issues.
- Communicate and train personnel, conduct drills at least once a year, evaluate emergency response performance, and initiate follow-up and corrective actions as appropriate.

Plan and program elements may often require the inclusion of and compliance with the regulations of congressional acts (e.g., CWA, CAA, CERCLA) for environmental impact events resulting in soil, water, and air contamination.

Response Plans for Environmental Hazards

In the event of hazardous chemical release (e.g., fire, explosion), employers must instruct contractors and subcontractors of the emergency response procedures in place. This communication must be accompanied by the release of type, locations, and potential of the threat and include information covering all other associated threats.

Events including fires, explosions, or chemical releases require the following actions:

- Notify the appropriate emergency response personnel, perform shutdown activities, and evacuate the immediate or affected area.
- Account for personnel within the designated assembly area(s).
- Assess the site for the potential of extensive evacuation, and evacuate if required.
- Conduct incident notification, reporting, and investigation according to policy/procedures.

Emergency Response System

The emergency response system is the process by which response teams react in an emergency. The incident command system is a standardized, hierarchal approach to emergency response.

The key concepts of incident command are unit of command; common terminology; management by objective; modular organization; and span of control.

A single unit of command to which all agencies report ensures that multiple agencies can operate as one team under the same leadership. Common terminology ensures agencies can communicate effectively. Management by objective ensures that teams manage time and resources to accomplish specific objectives by priority. Modular organization groups agencies based on their roles so that only the required agencies respond to an emergency based on the scope of the incident, type of hazards, and available resources. The span of control limits the duties and responsibilities under the control of each agency.

A crisis features three common elements: the element of surprise, the threat posed to the organization, and the time in which to act. Crisis management is the process by which an organization responds before, during, and after a crisis.

Emergency response equipment includes first aid kits, bandages, flashlights, eye wash stations, batteries, and other items necessary to respond to an emergency.

Media refers to the medium used to communicate in an emergency. This includes internal communication systems, social media, text messages, television, and other media, depending on the scope of the emergency.

Emergencies on construction sites can involve many types of materials and equipment. Examples include flammable liquids, corrosive chemicals, mechanized and motorized equipment, and even members of the general public, if they are not prevented from accessing the site.

Advanced emergency planning is the primary consideration when preparing for a potential disaster. Common emergency preparation plans have action guides or checklists, response plans, emergency management plans, and mutual aid plans. OSHA's minimum requirements for an employee emergency action plan (EAP) in construction can be found in 1926, subpart C, General Safety and Health Provisions.

In an emergency, the first priority is to safeguard personnel, equipment, and processes, in that order. Safeguarding the health and safety of personnel must always take precedence over equipment, tools, machinery, and all other assets.

Escape procedures and route assignments must be properly delineated; everyone on the site must know what to do and where to go during an emergency situation. In selecting an assembly (or muster) point during a site evacuation, a CHST should consider all factors and specific site conditions so that employees will not inadvertently be placed in further danger when proceeding to their assembly point.

At larger sites, designating multiple assembly points will allow employees to proceed to the nearest one when the alarm is sounded; this is especially advantageous when employees will be working at multiple places on the site during the workday. Assembly points should not automatically be fixed for the duration of the project; the assembly locations may need to change during the course of work and whenever any situations require a change, although this may not be necessary at smaller sites and for projects completing in a few days.

Whoever is selecting the assembly point should be familiar with the general geography of the site. This should include elevation, terrain features, landmarks, approximate distances, roads, and travel paths for vehicles and mobile equipment, along with any obstacles and barriers.

For sites involving the potential for a poisonous gas exposure like hydrogen sulfide (H_2S) (for example, oil field construction service and support operations), further considerations like wind conditions must be noted so that the assembly point is in the safest place possible.

The EAP will also need to describe how employees will be accounted for following an evacuation to an assembly point. The procedure can be as simple as a head count and writing it down in a notebook or having assembled employees write their names on a sign-in sheet like the type frequently used at the beginning of a workday; it does not need to be complex. Whichever type of procedure is used, it will need to be addressed in the EAP, and employees must be made aware of it, as they must be familiar with all provisions of the EAP.

Designated first-aid responders and all other medical response duties must also be addressed in the EAP. This is a formal and very important requirement that will ensure that only qualified personnel are delegated the authority to undertake these tasks; injuries have been inadvertently exacerbated by first-aid actions taken by well-intentioned but insufficiently trained first-aid responders.

Reporting procedures and designated contacts are also required to be in an EAP. Examples of common contacts include local fire and police departments, county emergency management personnel, regional poison control centers, and other relevant parties such as the hiring client and landowners.

A CHST will need to ensure that all emergency response provisions and responsibilities are communicated to employees. While OSHA does not require documentation of this, it is advised that employees acknowledge the receipt of this information, such as with an emergency orientation checklist or a simple sign-in sheet.

First Aid or Medical Needs

The successful implementation of a sound construction health and safety program will minimize the number of first-aid injury events. But even with the best planning, incidents do happen, and there will be cases when first aid must be promptly and correctly administered.

Because a serious injury, such as a fall from heights or electrocution, is more likely on a construction work site than in a general office location, personnel qualified in basic first aid must be available when prompt emergency medical services are not immediately available.

OSHA's requirement for first-aid availability at a construction site is that when emergency responders cannot arrive within 3 or 4 minutes, there must be at least one person on the site who is qualified to administer basic first aid until medical personnel with advanced training can arrive. The timely administration of basic first aid within the first few minutes of an injury has saved many lives in the construction industry.

First-aid training must be documented to satisfy OSHA's requirement for the training. OSHA generally does not require training documentation, but first aid is an exception. First-aid training is acceptable to OSHA if equivalent to Red Cross or Bureau of Mines curricula and is readily available in many locations. Many first-aid training instructors also provide cardiopulmonary resuscitation (CPR) and automated external defibrillator (AED) training, and these topics can all be addressed in a single course.

First-aid kits on construction sites must be placed in weatherproof containers, and each type of item in the kits must be individually sealed. The contents of first-aid kits must be inspected prior to assignment to a site and at least once per week on-site to ensure that spent items are replenished as needed.

Universal Precautions

OSHA's blood-borne pathogens standard requires that employees with a reasonable anticipation of exposure to blood or any other potentially infectious materials receive specific, annual training in blood-borne pathogens. Other potentially infectious materials include bodily fluid such as semen, vaginal secretions, pleural and pericardial fluids, amniotic fluid, and any bodily fluid that is visibly contaminated with blood. Since it is difficult to differentiate between bodily fluids that are potentially infectious and those that are not, many organizations will employ **universal precautions,** which treat all blood and fluids as if they are infectious.

Training topics must include identification and knowledge of blood-borne pathogen hazards, control methods, emergency actions, personal protective equipment (PPE), the hepatitis B vaccine, the employer's exposure control plan, and general housekeeping considerations. OSHA generally does not require training documentation, but blood-borne pathogens training is an exception, and training records must be retained for at least 3 years from the time of training.

OSHA's blood-borne pathogens requirements are found in 1910, subpart Z. While a General Industry Standard area, these requirements also apply to employees in the construction industry when there is a reasonable possibility of an exposure, such as designated first-aid responders, occupational health nurses, field medics, and any other personnel who are specifically designated to respond to injury events demanding immediate attention on a construction site. However, this does not include all employees on a construction site; dealing with blood should not be considered to be within the realm of normal experience for most construction employees, since no site should experience so many injuries of this severity that everyone needs to be trained in first aid and, consequently, in blood-borne pathogens.

Some construction project managers will propose sending all employees at the project site to first-aid training to ensure there will always be sufficient means to administer basic first aid in the event of a medical emergency. However, this is not necessarily the most ideal course of action in every circumstance.

When making a recommendation as to how many employees on a construction site will receive first-aid training and be designated as basic responders, a CHST should determine how many will be sufficient for a worst-case scenario while also taking into consideration the total cost of first-aid training and also blood-borne pathogens training, since that will also be required of all designated first-aid personnel. If a site will have thirty employees, and having four of them trained in basic first aid will more than cover any anticipated emergencies as well as provide for coverage when other first-aid responders are on vacations or otherwise away from the site, etc., then it may not be in the contractor's financial interests to send all thirty to first-aid along with blood-borne pathogens training.

A CHST must always consider the financial aspects of any decision involved in construction hazard control. First-aid and blood-borne pathogens training will involve course fees, paying employees while they are in training and not constructing, and also a much greater amount of training documentation and recordkeeping duties, which will take time and cost more. If the CHST has analyzed the situation and determined that having thirty employees trained in basic first aid will not make the site any safer than when four are trained, then this should be addressed with management so a well-informed decision is made. However, there are facilities and hiring clients who will require that all personnel on their property receive first-aid training. In that instance, the decision is already made.

Employers must make a hepatitis B vaccine available to all employees with job duties involving a reasonable exposure to blood and other potentially infectious materials, such as first-aid responders. OSHA does not require that employees receive the vaccine, but employers must make it available and must pay for it when employees elect to receive the vaccination.

An exposure control plan must be developed when one or more employees have an occupational exposure to potentially infectious materials. The plan must contain an exposure determination including a list of all job classifications involving a potential exposure, which in the construction industry will mostly apply to first-aid responders and other on-site medical personnel. The plan must be reviewed annually to determine if updates are necessary, and it must be made available to all employees for review.

Employers must provide adequate PPE for employees who will be in contact with potentially infectious materials. Examples include examination-style gloves and eye and face protection. Whichever the type of PPE, it must prevent potentially infectious materials from passing through and contacting employees' skin, eyes, mouths, and any mucous membranes, and also their clothing under the PPE.

PPE must be removed before leaving the work area and placed in a designated container for decontamination or disposal. Most of the PPE used for first-aid purposes in construction will be of the inexpensive and disposable type, so disposal is common in construction. Any contaminated equipment or tools must be decontaminated prior to being placed back into service.

Plan for Emergencies

Organizations must develop a written emergency plan for potential emergency situations and disasters. Emergency plans must provide protection to employees, the public, and the environment. The best way to maintain health and safety in the event of an emergency is to conduct advanced planning according to project specification, conditions, and geographic area.

Planning for emergencies should be based on priority with respect to events most likely to occur and the potential severity of impact. With respect to naturally occurring events such as floods and earthquakes, organizations should conduct extensive research to determine what threats are most likely to occur within prospective project areas. To reduce organizational pressures, emergency and disaster services can be contracted to appropriate agencies. The following list includes manmade and natural events that organizations must consider when planning for emergencies:

- Fires and explosions
- Hurricanes and tornadoes
- Earthquakes
- Floods
- Disbursement of toxic chemicals in the atmosphere
- Terrorism and workplace violence

When writing emergency plans, organizations should seek cooperation with any neighboring entities and public organizations. Written plans must cover a multitude of elements including the following:

- Handbooks and other materials related to emergency procedures that are available to employees (e.g., facility layout, evacuation routes and assembly areas, locations of response/first aid equipment)
- List of potential/anticipated emergency situations and risk statements
- List of safety process elements such as incident investigations and hazard mitigation and controls
- Description of communication equipment and methods with contingencies
- Description of alarms and warnings
- List of cooperative measures and communications with local firefighters, law enforcement, utilities, and emergency medical services and first responders
- Description of performance elements such as drills (fire and evacuation) and training for fire extinguishers, medical and first-aid procedures, hazardous materials spill response, and rescue

A chain of command should be established for communication and reaction to a given emergency situation selected by the organization's advisory committee and/or emergency coordinator. Personnel should be selected based on their capacity to perform under stress, rather than their organizational title. The chain of command should be kept as small as practical for speed and efficiency in emergency situations.

Safety Program Development and Implementation

Health and Safety Standards

In an effort to implement a national standard occupational safety and health program, the U.S. Department of Labor established the OSHA of 1970. Prior to OSHA, states established and managed their own occupational safety and health programs. These programs varied from state to state, with some more strict and some less strict in their monitoring and enforcement of safety programs. While some employers neglected to establish a safety program or ran a poorly managed one, other employers voluntarily invested significant resources to establish and maintain an occupational safety program.

In order to equalize the perceived economic advantage held by nonparticipant business owners and corporations, the government created OSHA to establish national minimum occupational safety standards for all business entities to share equal accountability for the safety of their employees (although states still typically run their own workers' compensation programs and OSHA departments). OSHA operates in concert with other regulatory agencies. The OSHA is responsible for the establishment and enforcement of mandatory health standards. The National Institute for Occupational Safety and Health (NIOSH) is the research arm of OSHA and enjoys the same workplace entry privileges as OSHA. The Occupational Safety and Health Review Commission (OSHRC) is the resolution arm of OSHA and will attempt to resolve citation disputes.

OSHA has a right to conduct unannounced site inspections but will do this with management consent or warrant. Site inspections are the primary method of enforcement utilized by the agency.

The five inspection methods include general schedule, complaint, fatality/catastrophe, imminent danger, and follow-up. General schedule inspections are conducted in high-hazard environments. Complaint inspections are generated internally and are filed confidentially by the employee or employees' representatives. Fatality/catastrophe inspections are conducted when a fatality occurs or when three or more personnel are hospitalized. Imminent danger inspections are instigated by any individual or party declaring such a scenario. Follow-up inspections tend to be associated with a past infraction or citation, and the subsequent inspection serves to verify infractions or cited items have been effectively mitigated.

The implementation of a worksite safety program must necessarily revolve around the baseline or initial site risk assessment whose main purpose is to build and document a base of reference for known and potential hazards unique to the worksite.

This initial audit provides the foundation for an active and continuous hazard identification program that includes daily safety walks (DSWs) and scheduled or impromptu inspections. Worker participation, feedback, and "buy in to the worksite" safety programs will go a long way toward perpetuating a sound safety culture.

The steps following hazard identification include the control measures applied to assess and mitigate such hazards. As stated earlier, the hierarchy of control measures includes elimination, substitution, engineering, administrative, and PPE. There is a possibility that a control measure for a hazard may be changed, should a variable present itself in circumstances, location, equipment, or materials. Documented evidence must reflect any worksite changes using a management of change (MOC) procedure.

The hazard identification program will be an ongoing process that manifests itself in monthly inspections, weekly inspections, and DSWs. As important as scheduled inspections and worksite monitoring for hazards are, nurturing an atmosphere of trust and common interest by the CHST in the workplace is of equal importance, as this encourages personnel to participate and provide hazard identification feedback.

The success of a worksite safety, health, and environmental program depends on some of the following factors:

- The installment and maintenance of a post-comprehensive hazard survey identification program
- Occupational safety and health surveys—worksite surveys include hazard identification, material inventory including hazardous chemicals (SDS), hazard communications and notices, air sample analysis, noise-level surveys, and ergonomic surveys
- The CHST must be aware of changes to materials, equipment, methods, and simultaneous operations that divert from the normal standard operating procedures (SOP) in place. The CHST must reconcile any potential MOC scenario with the requisite hazard analysis and mitigation measures including the documentary evidence relative to the action. This includes CAPA plans. All hazards must have related CAPA plans with each identifying root cause and a mitigating solution or solutions assigned to personnel with a required deadline for plan implementation. All CAPA plans must be approved by project management in accordance with regulatory bodies and regulations. Any CAPA not corrected will put the project at risk for further incidents and noncompliance.

A sound knowledge of applicable standards and safety legislation and regulations (and the capability to source pertinent reference documents) for the worksite is the responsibility of the CHST.

Employers are responsible for providing a safe workplace to employees. The safety professional is responsible for maintaining knowledge of regulations and how they apply. Science and technology can impact safety methods, best practices, and other elements, which can cause current standards and codes to change. Researchers should understand that the Internet contains vast amounts of inaccurate information and nonfactual elements based on assumptions and biases. Therefore, researchers must be diligent in efforts regarding cross-referencing, fact-checking, and confirming the credibility of sources.

Enforcement includes an array of subtopics such as inspections, high-hazard alerts, penalties and enforcement for severe violators, and letters of interpretation. Training and Education includes e-tools that provide web-based interactive training to users and safety and health topics that provide information on detecting and controlling hazards. Both are designed as training tools and provide relevant OSHA standards, how the standards apply, and information compliance.

Safety and health professionals can also reference OSHA's frequently asked questions (FAQs) to gain further understanding of regulations and how they apply. OSHA FAQs address over thirty safety- and health-related topics such as inspections, falls, incident reporting, training, etc. Each topic subsection provides a wealth of information and supporting regulations helpful to researchers.

Effective communication is the primary conduit for worksite safety, and the CHST is accountable for the proper and effective dissemination of topical worksite safety, health, and environmental information. The CHST will utilize the following communication methods: industry standards publications; regulatory resources and publications; checklists; in-house and contractor Health, Safety, Security, and

Environmental (HSSE) plans; HSSE signage; posted notices; and most importantly, job hazard analysis (JHA), worksite presence, and conversational engagement with all levels of personnel.

The CHST must approach this challenge with enthusiasm, impartiality, and empathy. The CHST will participate in the prioritization of worksite hazards, mentoring of worksite personnel, and corrective actions and resolution of any resultant worksite conflicts.

Best Practices

The foundation of effective safety is based on an in-place culture of organizational safety. This can be described as all levels of an organization holding common beliefs for the prevention of injury and illness. Organizations have the duty of providing a safe and healthy workplace to employees or, as stated by the General Duty Clause, a workplace that is "free from recognized hazards causing or likely to cause death or serious physical harm." This mechanism ensures the input of critical elements for site safety plans, processes, and performance.

Organizations can reference OSHA 3071 to:

- Identify workplace hazards
- Understand risk
- Control common hazards of the workplace
- Obtain information relating to selecting JSA/job hazard analysis (JHA) criteria and developing and using JSA/JHA

Organizations can reference model organizations that are part of OSHA's Voluntary Protection Program (VPP) to enhance existing programs and workplace safety culture. Section 3071 also outlines consulting services that are available to companies for addressing existing hazards and violations.

Additional information covers OSHA-approved state plans. While some states are compliant to federal standards, others maintain their own standards that must meet or exceed those of OSHA. Priority in complying with applicable standards and codes is often a matter of territorial jurisdiction.

Safety Preplanning—Best Practices

- Identify project hazards early in the design stage using formalized hazard analysis techniques.
- Ensure Environmental Health and Safety (EHS)–related processes, including hazard/risk and associated data, are supported by documentation and are part of contract documents.
- Incorporate the results of the hazard identification/analysis process into various design reviews of the project.
- Ensure EHS processes are continuously active/ongoing throughout the project life cycle.
- Describe and document a summary of all hazards, associated controls, and performance data throughout the project.
- Ensure site security issues are addressed and based on the scope of the project and are part of the safety plan.

Effective management of employer contracts is vital for providing safety to employees. Contract regulations most relevant to construction are located within 29 CFR 1926. However, construction employers must also adhere to sections contained in General Industry (CFR 1910). Employers can reference information from the American Institute of Architects (AIA) for contract development, formatting, and many other considerations for drafting effective contract documents.

OSHA standards outline the general guidelines for responsibilities and coordination among employers (owners), contractors, and subcontractors. Each party must understand their respective roles and responsibilities and how they apply. Applicable standards for construction contracts and multi-employer work sites are found within CFR 1926, subparts A and B. Most notably, this section mandates that all employers involved in work must ensure that no employees are exposed to hazardous work conditions, and that this duty cannot be contracted out or delegated to another party. Standard interpretations and letters of reply provide further understanding of these standards. Other applicable regulations are contained in 48 CFR, subpart 52, for contracts and construction site conditions.

Best Practices—Participants

Owners /Host Employers:

- Practice diligence with respect to selecting contractors with good safety records.
- Place focus on monitoring and maintaining scheduling and coordination, along with work/contractual relationships throughout the phases of a project.
- Ensure that safety is a joint responsibility. Worker safety should be the responsibility of the respective employers as contractors and subcontractors. Reliance on contractors and subcontractors should be expressed within the contract.
- Avoid attempting to assume control of contractor/subcontractor safety processes, as responsibility for safety is shifted contractually.
- Ensure clarity and understanding of contract language.

Contractors:

- Avoid absorbing the entirety of safety responsibilities across the job site. This should be done through ensuring subcontractor provision of safety expertise and assumption of responsibility with respect to their scope of work on the site.
- Express safety as a high value by holding meetings and encouraging feedback.
- Require the sharing of safety practices and related information among participants (contractors/owners and contractor/subcontractors).
- Request the immediate reporting of safety violations, near-miss incidents, and other concerns whereby subcontractors are not deemed responsible.
- Ensure contracts include clauses relating to unforeseen conditions (differing site conditions [DSCs]).
- Ensure clarity and understanding of language in all contracts.

Subcontractors:

- Avoid assuming responsibility for safety beyond what is stated as required by the contract.
- Ensure written clearance of any responsibilities they did not create or of which they were not made aware.
- Acknowledge reliance on the expertise of contractors and other subcontractors.
- Report incidents outside their scope of responsibility to the contractor so they may be addressed.
- Instruct employees to remain in areas designated for their work as assigned by the contract.
- Ensure clarity and understanding of language in all contracts.

Best Practices: Contract/Project Life Cycle

Prequalification

Exercise diligence in vetting contractors' safety performance and the level of risk they present. Host employers should review components of contractors' performance such as:

- Pre-task planning
- Environmental Health and Safety (EHS) policies
- Safety meetings
- Safety statistics
- Training
- Inspections
- Incident investigations and reporting

Pre-task Risk Assessment and Preplanning

This phase should involve detecting hazards and exploring methods to eliminate and mitigate hazards. Proactive processes such as job safety analysis (JSA) and failure analysis methods should be incorporated into this phase prior to beginning work activities.

Contractor Training and Orientation

Contractor and subcontractor employees should be introduced to site-specific processes. The primary processes for which employees receive training include permit-required activities such as hot work like soldering and welding in which the risk of igniting flammable material exists, confined-space work, LOTO procedures, and PPE. Further measures to complete this phase include OSHA course completion and evaluation of the worker skills that are necessary for the safe performance of processes.

Safety management practices within the project monitoring phase include the monitoring and tracking of processes such as:

- Submitting safety observations
- Reporting unsafe acts and conditions
- Reporting noncompliance
- Reporting incidents
- Handling lost-time injuries
- Handling material and financial losses
- Reviewing workers' performance (performed by contractors)

Post-Project Evaluation

This phase is useful for employers to evaluate whether a contractor qualifies for further work. During this phase, employers assess how a contractor performed with respect to:

- Overall safety statistics
- Customer service satisfaction
- The quality of finished work
- Scheduling
- Change management

Employers assess and retain the information they gained during the vetting process and evaluate how well contractors performed and met the expectations outlined in the contract.

Site-Specific Safety Plans

For any new construction project, management planning encompasses tasks such as:

- Identifying the necessary materials and equipment for the project
- Assessing the appropriate number and type of workers needed to complete the project
- Creating the project schedule
- Tallying the associated costs

For good safety performance throughout the project's life cycle, project planners apply risk assessment principles to detect losses associated with compromised workers, public safety, and materials. These elements are included in the planning stage.

Effective safety planning requires owners to leave construction activities in the hands of contractors who can more effectively orchestrate project safety. This reduces ownership liability, demonstrates top-down leadership, and places a fundamental focus on safety through design.

Planning begins with the cultural assessment at all levels, including program and project managers, contractors and subcontractors, and all associated workers. From a safety standpoint, the main goal is to maintain an injury-free workplace through a culture that pursues safe work practices and emphasizes both individual and team responsibility. The following illustrates some of the many essential elements for effective site-specific safety planning.

Multidisciplinary Teams

Multidisciplinary teams distribute safety considerations among various personnel and their respective functions to acquire a broadened scope of feedback. When the scope of safety broadens beyond one or a few individuals, performance becomes far more effective. Under this condition, coordination among managers, architects, engineers, and field employees brings diverse input and feedback into the planning phase to help ensure safety solutions are consistent with scheduling, quality, and budgetary and design goals. In turn, these elements are also consistent with the effective provision and goals of safety. Employers should hold regularly scheduled meetings to keep personnel informed and to welcome input about the project. This is an essential part of creating and maintaining site safety.

Preconstruction Meetings and Toolbox Talks

Contractors should meet to communicate the general safety requirements of the job site. Subsequently, contractors should hold meetings with subcontractors to educate them of these same requirements. Competent individuals on-site should be selected as safety representatives for their respective contractors and subcontractors. Either these individuals or site supervisors should attend regular preconstruction meetings to review safety elements specific to the site.

Pre-task meetings (toolbox talks) between foremen and crew allow discussion of upcoming work and the safety considerations of the shift. To maintain safety as a major focus, managers should meet periodically to ensure an appropriate and effective balance between safety and production is maintained throughout the project.

Measurements and Record Keeping

Employers and contractors must track and measure safety performance. A critical function of program management and continual improvement is identifying trends and indicators relative to safety warnings issued, loss, injury, and near-miss incidents.

Recordkeeping includes the documentation and filing of the training that was delivered, the inspections and audits that were conducted, the environmental hazards that were monitored, the drug screenings that were performed, etc.

Training

Effective safety planning must include a continually active training process. Safety training consists of various processes and procedures and is completed for the primary purpose of avoiding injury and illness. Training must be performed prior to the task and documented. At all levels, focus must be placed on the alignment of a clear message of the expected procedures to maintain safety and a high quality of work. This reduces misunderstandings and promotes safety awareness.

Job Safety Analysis (JSA)

A JSA is a process that examines the procedures used to perform a task in order to detect observable and foreseeable hazards for which elimination/mitigation measures can be implemented. Corrective actions often include process or procedural alterations. These actions should be completed in the order of hazard priority by applying hierarchical concepts with regard to controls. As this process is proactive in nature, it is an effective part of preplanning. JSAs can also be used for training and retraining purposes, as necessary.

Job Progress Meetings

Job site meetings should occur in order to discuss any issues related to:

- Safety coordination
- Performance reviews
- Problem-solving
- Events to come

This forum should welcome employee suggestions and demonstrate management commitment and responsiveness.

Construction Means, Methods, Equipment, and Materials

To ensure effective safety performance, organizations must fully comprehend the entire scope of construction activities that will occur during the project. This understanding is critical for site safety preplanning. Knowledge of activities and processes will aid the participants in anticipating the potential system deficiencies that will require improvements as well as the hazards that can be eliminated or mitigated before the project begins. This is vital to ensure project safety processes align with other components such as material selection, scheduling, and the work to be done.

As part of preplanning and follow-up, a safety practitioner must question components involving performance and compliance for each construction process that will comprise the project. This questioning process can be designed as an abilities assessment form to obtain "yes" or "no" answers or as a checklist for rating performance criteria. These assessments or checklists help determine what improvements should be made before construction begins. The following outlines some of the many construction processes (methods, equipment, materials), with examples of respective performance/compliance elements.

Fall Protection

- Does the design reduce the number of stairs and ladders needed?
- Has fall-prevention equipment been considered before fall-protection devices?
- Is the system designed with or to receive anchor points and fall-restraint devices?
- Were considerations given for providing fixed ladders and work platforms for work at heights?

Lockout/Tagout—Hazardous Energy Control

- Has controlling the release of hazardous energy been considered as part of the design?
- Do lockout/tagout (LOTO) procedures exist?
- Have the prospective energy forms been identified?
- Are LOTO devices accessible, standardized, and in the required condition?

Electrical Safety

- Does the electrical system meet all applicable code requirements?
- Have grounding procedures for tools and equipment been considered?
- Is high-voltage equipment isolated as required?
- Are components adequate where combustible vapors may exist?

Walking and Working Surfaces

- Will walking surfaces be designed of or designed to incorporate nonslip materials?
- Does the design allow for walkways free of obstruction?
- Do all floor and wall openings meet OSHA requirements?
- Will utilities be directed away from walking surfaces?

Confined Spaces

- Has confined-space work been eliminated or reduced in project design?
- Have items deemed as confined spaces been assessed for permit requirements?
- Has the potential for introduced hazards been assessed with confined-space work?
- Have monitoring for hazardous atmospheres and other confined-space safety requirements been considered?

Worksite Assessment and Audit

A construction health and safety technician (CHST) has a wide array of practical and regulatory guidance at his or her disposal with which to perform a well-conceived and thorough audit, identifying potential hazards. All pertinent standards, codes, and applicable best practices regarding health, physical, construction, and environmental safety must be referenced in documentation, which cites applicable laws and regulations. These may include international, federal, and state laws and regulations along with site-specific, contractor, and subcontractor Health, Safety, and Environment (HSE) plans. A CHST must be familiar with safety program management, Occupational Safety and Health Administration (OSHA) regulations and inspections, the basic tenets of hazard communication, on-site job hazard analysis (JHA)/mitigation, accident investigation, and documentation practices as required by all applicable laws.

It is the CHST's duty to arrive on-site and carry out an audit in a manner illustrating sound documentary and practical procedures. The CHST must not necessarily memorize the myriad of mandated procedures and rules governing the worksite, but he or she must know how to source and reference all applicable documentation for the task at hand. The purpose of an on-site audit is to identify any and all existing and potential hazards. From there, it is the CHST's role to recommend corrective actions to ensure the worksite comes into full compliance with all applicable laws and regulations. The CHST must be familiar with both those regulations and best practices in order to provide sound corrective actions. The CHST may then participate in repeated inspections in accordance with any site-specific corrective action plans in order to ensure the actions have been appropriately documented, implemented, and resolved.

While it is not necessary for a CHST to commit all the applicable international, federal, and state laws and regulations to memory, please note that for the CHST exam, a test taker must be intimately familiar with required reference materials prior to taking the test. A key to success in CHST certification and the candidate's subsequent ability to perform effective worksite audits is having a familiar knowledge base from which to work. A CHST who is unaware of basic rules and regulations will not be able to perform thorough audits and recommend appropriate corrective actions.

Safety Management

Effective safety performance requires that safety and health roles and responsibilities are shared and practiced by all. Roles and responsibilities should be clearly defined and specified within safety and health programs, policies, procedures, and documents associated with contracted work.

Owners and upper-level management are responsible for developing and procuring safety and health programs and policies and further defining safety roles and responsibilities. This should be done with the assistance of a joint safety and health committee and/or the combined efforts of safety professionals, supervisors, mid-level managers, engineers, and department representatives. Programs and policies should garner support and mutual agreement among members of upper management. Management must also demonstrate commitment to safety and make policies and associated safety content available to employees.

The safety professional is responsible for implementing safety processes throughout a project's life cycle and, most importantly, is the driving force for integrating safety into the design stage of projects. Safety personnel are responsible for practicing effective communication with all levels of an organization. Communicating with top management is critical for the safety professional to express the importance of maintaining and improving the programs and policies in place as well as for management to demonstrate commitment to these elements for the continued protection of workers. Communicating with workers is critical for ensuring safety processes and procedures are effective and aligned with work activities.

Employees (workers) are most often the backbone for safety success. Well-trained and educated workers play a large part in detecting potential concerns. Workers can determine if safety and health policies and procedures align with the established processes or if policies and procedures consider injury and illness, quality, and production. Employees are also responsible for familiarizing themselves with safety and health policies set forth in the form of postings, handbooks, procedures, etc. Other employee responsibilities include reporting any injuries and incidents, conducting inspections as assigned, and using personal protective equipment (PPE) as provided and prescribed by the employer.

Equipment Inspection

Scheduled inspections and preventive maintenance procedures detect and address equipment deficiencies and potential hazards before an incident occurs. This provides a higher level of safety to employees and positively impacts quality, production, and profit. Inspections are performed by personnel such as managers, supervisors, and maintenance workers and should be led and monitored by the safety professional. Inspection plans/programs are most effective when a combination of daily and scheduled inspection processes is used and properly enforced.

Organizations should emphasize the importance of routinely inspecting machines and equipment. The purpose of these inspections is to identify potential hazards and correct them through preventive maintenance before an incident occurs. Beyond improving employee safety and providing benefits to quality and production, inspections can positively impact performance, efficiency, and profitability. They also can be performed by different members of an organization, including managers, supervisors, maintenance, and medical personnel, and they can be led and continuously monitored by the safety professional. Inspections can be done on a **continuous** (daily) basis or on a **scheduled** (interval) basis. Inspection programs are most effective when using a combination of both.

Prior to implementing an inspection program, an organization should determine the following:

- What requires an inspection?
- Which parts should be examined?
- The appropriate physical condition of items for inspection
- The frequency of inspection
- Which personnel will conduct the inspection?

Basic Risk Management Concepts

Workplaces, especially construction sites, present a certain amount of risk from injury, exposure to chemicals, budget shortfalls, missed deadlines, et cetera. The first step in mitigating risk to avoid claims is to identify and assess potential risks to better plan and manage them.

Mitigating risk means reducing negative impact by avoiding, limiting, transferring, or accepting the risk. The process of mitigating risk includes identifying hazards in the workplace that could produce negative effects; assessing the likelihood and impact of the risk in the event of exposure; identifying ways to limit exposure or impact; removing hazards that present risks or implementing strategies to control exposure; and devising fallback plans or contingencies if exposure is unavoidable.

Depending on the type of industry or the nature of the project, hazards on the job site may present a risk to public safety. Risk management should include strategies to communicate with the public in case of exposure and respond efficiently to limit negative impact and protect the safety of the public.

Builder's risk insurance covers damage to property that is under construction. This policy protects the builder's interest while they are working on a property to ensure they are not liable for losses caused by natural disasters, fire, or high winds. It also ensures the builder is not liable for damage that occurs during the course of the construction work. It does not cover flood or earthquakes. Builder's risk coverage ends when the construction is completed. Such a policy can be purchased by the contractor (builder), the property owner, or both.

Builder's risk coverage applies to the property, as well as the equipment and the materials used to complete the construction work, but it does not cover injuries to workers or protect contractors or property owners from claims by workers or third parties if an injury occurs on the property during the construction work. General liability insurance covers these types of losses from accidents on the job or claims against the contractor or property owner if someone is injured by the completed work. General liability does not cover claims due to faulty workmanship or negligence.

Value engineering is an estimation method used to find the most cost-effective project construction processes without compromising worker safety; this makes the accuracy of the initial hazard identification and the ongoing hazard identification methods such as the JSA even more important. A JSA is a mini risk assessment that enables the workers charged with a task to participate in the entire safety vetting process while providing a daily sounding board and worker interaction. This is always a great opportunity for the CHST to mentor in the application of the key risk management elements: identify, assess, analyze, manage, monitor, and control. The CHST has the capability to inspire a meaningful exchange between workers during a JSA. To achieve this aim, the CHST must effectively encourage expressions and ideas from the most reticent workers while arbitrating the discussion to rein in the more outspoken ones.

To balance both of these elements, the CHST is obliged to listen more and to choose speaking points wisely and economically.

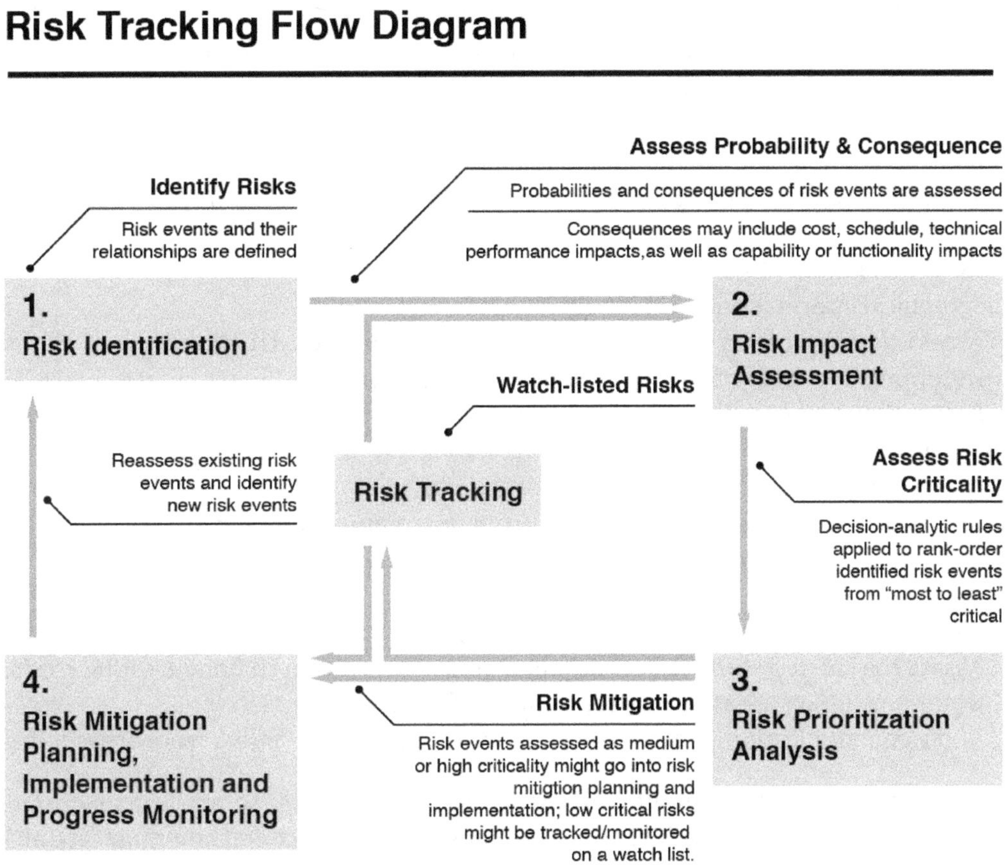

Construction Site Conditions and Safety

Assessing the general construction site is an important part of project design and planning. Evaluating site conditions should occur prior to planning and must include the surrounding areas. Site evaluation involves a three-step process. The initial step involves assessing site conditions and collecting relevant data. The second step involves data analysis crucial to making informed decisions for design plans. The third step requires the data be applied to the initial concepts and designs. Information can be collected from several categories. The data can then be analyzed and used in several ways including procuring cost-effective designs, proving site safety, and protecting the environment.

Some examples are as follows:

- Soil Information
    - Determine the capacity to support parking lots, streets, and footings.
    - Apply the slopes required in trenches and excavations.
    - Reveal the level of the water table and location and depth of bedrock.
    - Assess erodibility and drainage ability.
- Vegetation
    - Identify isolated areas that are effective in resisting soil erosion.
    - Assess the vegetation's ability to filter sediment and pollutants from storm water and runoff.
    - Determine areas beneficial for aesthetics.
- Topographical Information
    - Assess slope gradients critical for building designs and construction.
    - Determine the location and direction of drainage systems.
    - Assess flow and capacity of surface water and storm runoff.
- Hydrologic Data
    - Assess how water flows into and out of the site.
    - Assess potential on-site flow volumes from upstream watersheds.
    - Determine site drainage impacts regarding runoff and flooding of downstream water bodies.
- Highways and Utilities
    - Ensure no construction elements interfere with existing highways and roads.
    - Assess the capacity for site drainage and ground alteration to flood existing roads.
    - Determine utility areas to be kept clear and free of obstruction.
    - Apply communication strategies and controls to protect workers from areas of high activity.

Evaluating site conditions is critical for effective design planning, worker safety, and environmental health. Where contracted work is involved, organizations should detect and communicate all known site conditions to the contractor by including them in the contract and ensuring agreement. Contractors must practice diligence in evaluating site conditions and ensure knowledge of the contents of the contract, their responsibilities, and their legal protections.

Incident Investigations

Following an incident, the first fact-finding efforts will range from a generalized survey of the situation to a formal investigation. The formality and complexity of the action taken will generally depend on the severity of the incident; minor near-miss incidents will not require a formal investigation involving multiple parties, and a significant injury investigation must not be limited to a brief survey.

Whichever investigation technique is employed, it must be adequate to determine probable causes; the objective of any incident investigation is to prevent a similar incident from occurring. The technique must also be sufficient to satisfy the program and policy requirements of the contractor, hiring client, and any other relevant third parties. There may be multiple parties requiring information, and there will be times when their requirements and agendas are not universally mutual. A CHST will be a source of guidance during an incident investigation and must be sufficiently knowledgeable of the organizations involved in order to effectively communicate with them during the investigation.

Since the incident has already happened and a CHST cannot travel back in time to prevent it from happening, an investigation will always be a backward-looking effort employing deductive analysis to identify all factors that could have contributed to the event.

When beginning an incident investigation, the priorities are to arrive safely, size up the situation, care for the injured, and lastly, protect property.

The most immediate supervisor, such as a foreman or project manager, is the lead person in most routine incident investigations. Since a foreman can be expected to know the work area and personnel better than anyone, it is reasonable to assume that the foreman will also be the most appropriate party to identify the causal factors that may have led to the incident. For this reason, it is not advised that a CHST or safety administrator be the lead person in routine incident investigations, and also because the supervisor who exercises control over the operation should ultimately be the person responsible for correcting the causal factors in the incident.

During a routine incident investigation, the standard role of a safety manager is to assist the supervisor conducting the investigation through observation and technical guidance as required. However, the supervisor must be effective in this role; otherwise, the fact-finding effort may result in useless data. In such a situation, the CHST or safety manager will need to have a more hands-on role in the investigation; depending on the supervisor's deficiencies, the CHST may even need to conduct the investigation.

For incidents involving fatalities, very severe or multiple-injury scenarios, and very costly property damage, it is advised that senior management formally conduct the interview. These scenarios may also involve lawyers and other third parties such as liability and insurance professionals. For these types of high-gravity events, the CHST will frequently serve as a technical advisor during the course of the investigation.

Management deficiencies are unifying characteristics of root causes. Deficient management practices ultimately empower personnel to introduce the causal factors that make an incident possible, and there are frequently multiple causal factors in an incident. For example, a particular equipment malfunction may be identified as a causal factor in an incident that injured an employee, and the decision of senior management to purchase inferior and unsafe equipment may be identified as a root cause from which the causal factor is derived. Or perhaps the contractor's human resources department hired personnel who were unsuitable for their roles, and this practice resulted in the introduction of causal factors in the incident, such as employee errors, inadequate job and safety training, and other human factors.

Actions such as corrective/preventive measures, disciplinary proceedings, and informational meetings with third parties like safety committees or union representatives should only take place once useful factual data has been gathered; actions taken without factual data are counterproductive.

Although the facts may ultimately reveal that disciplinary measures are necessary, the initial analysis must be fact-based and not oriented toward finding fault for the purpose of determining appropriate disciplinary measures. A proper investigation identifies the root causes of the incident through fact-finding, not fault-finding.

Employee interviews should take place as soon as possible because important details may be forgotten, and the general clarity of information recall will degrade quickly in the first few days following an incident.

The best place to interview a witness is the incident site. The witness will be able to identify points of reference on the site and more fully explain the details involved. Moreover, communication errors are less likely to occur because the interviewer will not have to presume what the witness is saying about a specific site detail when both are present.

The incident site will not be an appropriate interview venue when the incident is very bloody or particularly ghastly in any regard because many employees will be upset by this and will not want to be on the site.

The next best interview place is a private and neutral location that will not intimidate the witness, such as a conference room. An administrator's office is not an advised interview location because it may put the witness on guard.

Witnesses must understand that the fact-finding interview is not a disciplinary interview, and so this should be communicated to them prior to asking any questions.

Handling Evidence

There is a general duty to preserve evidence whenever it can be reasonably foreseen that the information may be relevant to the investigation. Formal chain-of-custody procedures will help to ensure that evidence is handled appropriately and hold employees responsible for doing so.

The starting point for custody is when the evidence is documented for context, time, conditions, and anything else material to the particular evidence. For example, identifying photo orientation and lighting would be useful for photo evidence.

Relevant and correct documentation and notation are of the utmost importance because evidence may be reviewed by parties that have never visited the site of the incident, and the evidence and witness interview transcripts may be the only sources of information for them. Corrective and preventive action (CAPA) records are also critical, historical documents that should be readily available for reference.

All employees assuming custody should be required to sign a chain-of-custody document and should receive a receipt of the transfer. The number of employees handling the evidence should be kept to a minimum to reduce the likelihood of errors and mishaps in the transfer process.

Determining the Root Causes of Incidents

When attempting to identify cause-and-effect relationships for hazard mitigation, the cause can be misidentified or found to be an undesirable event with yet another cause. *Root cause analysis (RCA)* is a technique where the safety professional describes the negative effect or loss to define the incident cause. This technique requires complete evaluation of the event and the hazards inherent in the system, determining deviations from acceptable requirements and analyzing situations essential for the event to occur. To define an incident, a safety professional determines the root cause through a series of who, what, where, when, and how questions. This process can reveal a myriad of undesirable conditions that can be mitigated.

Decades ago, and with limited research, William Heinrich proposed that 88 percent of worker injuries were caused by workers' unsafe acts. Heinrich developed the **Domino Theory**, which is still used today. The Domino Theory is similar to the **Chain of Events Theory**. Both focus on workers' unsafe acts in a string of events that result in incidents and accidents with unsafe conditions as contributing factors. Following the review of events, the mitigation strategy relies on the removal and/or correction of the

key domino (weakest link) to break the chain reaction. Both theories also review events that can pinpoint deficiencies in worker instructions, pre-planning, programs, and policies. In addition to unsafe acts, these should be considered as potential "weak links" in a chain for correction.

Workers' safety levels and performance can be negatively impacted by a myriad of environmental factors. Although causes can stem from management deficiencies and unsafe conditions, other contributing factors can occur outside the workplace. These are characterized as **life events**. The **Adjustment Stress Theory** considers a multitude of life event situations that result in worker stress and depression (e.g., divorce, death of a loved one, or personal illness). Life events can negatively alter an employee's mental state, compromising their focus and awareness. Once introduced into the work environment, the worker can become an increased liability to themselves and others. Hazard assessment and mitigation can be performed through employee monitoring by the safety professional and management for subsequent counseling and/or treatment.

Incident Reporting and Follow-Up Procedures

To improve safety and health, all incidents must be reported. Incidents go unreported for several reasons. One reason is a safety program that includes monetary incentives for safety performance. In this case, an accident (and its associated report) results in the employee or department losing some or all of the quarterly cash/bonus due to the incident. Because of this, an employee may seek to avoid monetary loss, ridicule, or embarrassment and instead choose to hide the incident by failing to report it. Such chains of events render safety programs ineffective and counterintuitive with respect to good safety culture behaviors and the goals of continual improvement.

Incident reporting and investigation are an after-the-fact response to an event or accident. The primary purpose of reporting and investigation is to determine the root cause and prevent the incident from recurring. Part of awareness in reporting incidents is knowing what should be reported. Aside from events where injury and loss have occurred, close calls and near-miss incidents must be reported as well. Such situations (close calls) serve as indicators that an accident will eventually occur. Effective site safety plans should outline incident investigation procedures and what they entail. Any corrective actions implemented after the incident should be tracked, monitored, and measured to determine the level of effectiveness in preventing further incidents. Such actions are critical to continual improvement efforts. The four general steps for incident investigations are as follows:

Describe the Scene

- Guard the scene via equipment (e.g. cones, barriers) to prevent important materials from being disturbed or removed.
- Record facts such as the time, date, and location of the incident; name(s) of the injured; a description of the severity of the injury; and the names of witnesses.
- Take photos and/or videos of evidence at the scene and information that could be critical for analysis.

Data Collection

- Take statements and conduct interviews as soon as possible (utilizing recent memory to acquire more accurate information). Practice greater caution when interviewing the injured person(s).
- Review equipment and maintenance inspection programs, work processes and procedures, and operating manuals.
- Review records for similar incidents and follow-up actions.
- Review processes relevant to worker training, discipline, and enforcement records.

Determine the Root Cause

- Apply root cause analysis through a series of "*Why?*" questions methodology; proceeding with *if not*, or *if so,* then ... ("Why?" repeating) to determine the origin or cause.
- Apply root cause analysis to detect failures in procedures and apply actions to program and process deficiencies, not to assign personal fault.

Apply Corrective Actions

- Present adequate causes for corrective actions to upper management.
- Avoid presenting less significant causal factors to management (e.g., not using PPE).
- Target and modify program-level elements (e.g., equipment selection, procedures, maintenance programs) for corrective actions and implementations.
- Follow up on corrective actions with tracking and monitoring to determine effectiveness in preventing further incidents.

The incident report will consist of a summary and narrative. The narrative includes findings, conclusions, and recommendations for corrective action. The summary is the most important component of the report because it identifies the need for the report; the summary must be written in a manner that will encourage the reviewer to read the narrative. A CHST will encounter administrators who may not be in the habit of reading reports; an effective summary will increase the likelihood that the rest of the report will be reviewed.

Documentation of Occupational Injuries and Illnesses

Construction establishments with eleven or more employees (on any day of the calendar year) are required to record employee injuries and illnesses on the appropriate OSHA recordkeeping forms. OSHA's recordkeeping standard can be found at 29 CFR 1904.

For the purposes of OSHA recordkeeping, OSHA's definition of an employee includes all full-time, part-time, and seasonal personnel. Temporary workers, such as those provided by a staffing agency, are also treated as employees by this standard so long as they are supervised like any other employee. Contracted sole proprietors are not treated as employees; a contractor does not have to record an injury occurring to a sole proprietor unless the sole proprietor is being directly supervised, which would be unusual. The owners of the contracting company are also not considered employees for the purposes of OSHA recordkeeping.

A single construction site (with eleven or more employees) is required to perform site-specific OSHA recordkeeping tasks if the site will be in operation for one year or longer. For sites in operation for less than one year, the particular site is not obligated to perform site-specific OSHA recordkeeping duties, but injuries and illnesses will still need to be recorded by the employer's particular business establishment the employees are assigned to.

A general contractor does not record the injuries and illnesses of subcontractor employees. Sometimes, there may be confusion about this due to OSHA's multiemployer citation policy. This policy allows OSHA to issue citations to a general contractor because of the safety violations of a subcontractor. For example, if a project's roofing subcontractor is cited by OSHA for the failure to provide fall protection, then the general contractor managing the project may also be cited by OSHA for failing to provide fall protection, even though the general contractor's employees were not exposed to fall hazards. OSHA adopted this policy to hold general contractors more fully accountable for the unsafe actions of their subcontractors.

However, OSHA's multiemployer citation treatment does not apply to OSHA record keeping, and a CHST may need to clarify this when a subcontractor incident occurs. If a subcontractor employee experiences an injury due to a fall, then only the subcontractor is required to record it.

OSHA does not require every injury and illness to be recorded; only incidents that meet one or more of a predetermined list of severity or event criteria are required to be entered on the OSHA recordkeeping logs.

The majority of incidents in construction will not meet the various criteria to trigger OSHA recordkeeping. Although it is common for contractors to record incidents even when OSHA does not require it for internal hazard-control purposes, it is extremely important that only incidents required to be on the OSHA recordkeeping logs be entered on these logs. If incidents that do not meet OSHA's severity or event criteria are also recorded on the OSHA logs, then the contractor will be overreporting OSHA-recordable incident events. Overreporting incident events will artificially inflate the contractor's incident rates, and contractors have been disqualified from work and bids because of their incident rates.

Because of the importance of accurate incident rates and the data from which the rates are derived, a CHST must commit to memory the particular events that will automatically trigger OSHA recordkeeping and the various types of cases.

All deaths must be recorded, with the exception of "natural causes" and preexisting health conditions and events such as heart attacks. A medical examiner or other licensed medical professional will need to make the determination that the death was not work-related to avoid log entry.

The employer must record a case if an employee misses one or more days away from work or is assigned modified duty following the day of the incident. Modified duty includes work restrictions and transfers. Examples of modified duty are partial workdays, lifting restrictions, and not performing routine job functions. Transfer duty is whenever an employee is temporarily assigned to another position; for example, a construction laborer could be transferred to the office to help with clerical tasks while a back sprain heals.

The OSHA Form 301 is the Injury and Illness Incident Report. This form is used to record basic details of an incident, such as employee information, day and time of the incident, and the particulars of the incident. The 301 must be completed within seven calendar days of the incident.

OSHA Form 301 (Injury and Illness Incident Report)

must be completed within 7 calendar days of the incident

OSHA's Form 301

Injuries and Illnesses Incident Report

Attention: This form contains information relating to employee health and must be used in a manner that protects the confidentiality of employees to the extent possible while the information is being used for occupational safety and health purposes.

U.S. Department of Labor
Occupational Safety and Health Administration

Form approved OMB no.1218-0176

This *Injury and Illness Incident Report* is one of the first forms you must fill out when a recordable work-related injury or illness has occurred. Together with the *Log of the Work-Related Injuries and Illnesses* and the accompanying Summary, these forms help the employer and OSHA develop a picture of the extent and severity of work-related incidents.

Within 7 calendar days after you receive information that a recordable work-related injury or illness has occurred, you must fill out this form or an equivalent. Some state workers compensation, insurance, or other reports may be acceptable substitutes. To be considered an equivalent form, any substitute must contain all the information asked for on this form.

According to Public Law 91-596 and 29 CFR 1904, OSHA's recordkeeping rule, you must keep this form on file for 5 years following the year to which it pertains.

If you need additional copies of this form, you may photocopy and use as many as you need.

Completed by ____________________
Title ____________________
Phone ____________ Date ________

Information about the employee

1) Full Name

2) Street

City ____________ State ____ ZIP ________

3) Date of birth

4) Date hired

5) ☐ Male
☐ Female

Information about the physician or other health care professional

6) Name of physician or other health care professional

____

____

7) If treatment was given away from the worksite, where was it given?

Facility

Street

City ____________ State ____ ZIP ________

8) Was employee treated in an emergency room?

☐ Yes
☐ No

9) Was employee hospitalized overnight as an in-patient?

☐ Yes
☐ No

Information about the case

10) Case number from the Log ______ *(Transfer the case number from the Log after you recorded the case)*

11) Date of injury or illness ______

12) Time employee began work ______ AM/PM

13) Time of event ______ AM/PM ☐ Check if time cannot be determined

14) **What was the employee doing just before the incident occurred**

Describe the activity, as well as the tools, equipment or material of employee was using. Be specific. Examples: "climbing a ladder while carrying roofing materials"; "spraying chlorine from hand sprayer"; "daily computer key-entry"

15) **What happened?**

Tell us how the injury occurred. Examples: "When ladder slipped on wet floor, worker fell 20 feet."; "Worker was sprayed with chlorine when gasket broke during replacement"; "Worker developed soreness in wrist over time."

16) **What was the injury or illness?**

Tell us the part of the body that was affected and how it was affected; be more specific than "hurt", "pain", or "sore". Examples: "strained back"; "chemical burn hand"; "carpal tunnel syndrome"

17) **What object or substance directly harmed the employee?**

Examples: "concrete floor"; "chlorine"; "radial arm saw". If this question does not apply to the incident, leave it blank.

18) **If the employee died, when did death occur?** Date of death

Public reporting burden for this collection of information is estimated to average 22 minutes per response, including time to review the instructions, search and gather the data needed, and complete and review the collection of information. Persons are not required to respond to the collection of information unless it displays a currently valid OMB control number. If you have any comments about these estimates or any other aspects of this data collection, contact: US Department of Labor, OSHA Office of Statistical Analysis, Room N-3644, 200 Constitution Avenue, NW, Washington, DC 20210. Do not send the completed forms of this office.

The OSHA Form 300 is the Log of Work-Related Injuries and Illnesses. This is essentially a basic spreadsheet for logging OSHA-recordable injury and incident events. Data from the OSHA Form 301 that identifies the person affected and a brief description of what happened will provide the background for the OSHA 300 log entry.

OSHA Log of Work-Related Injuries and Illnesses

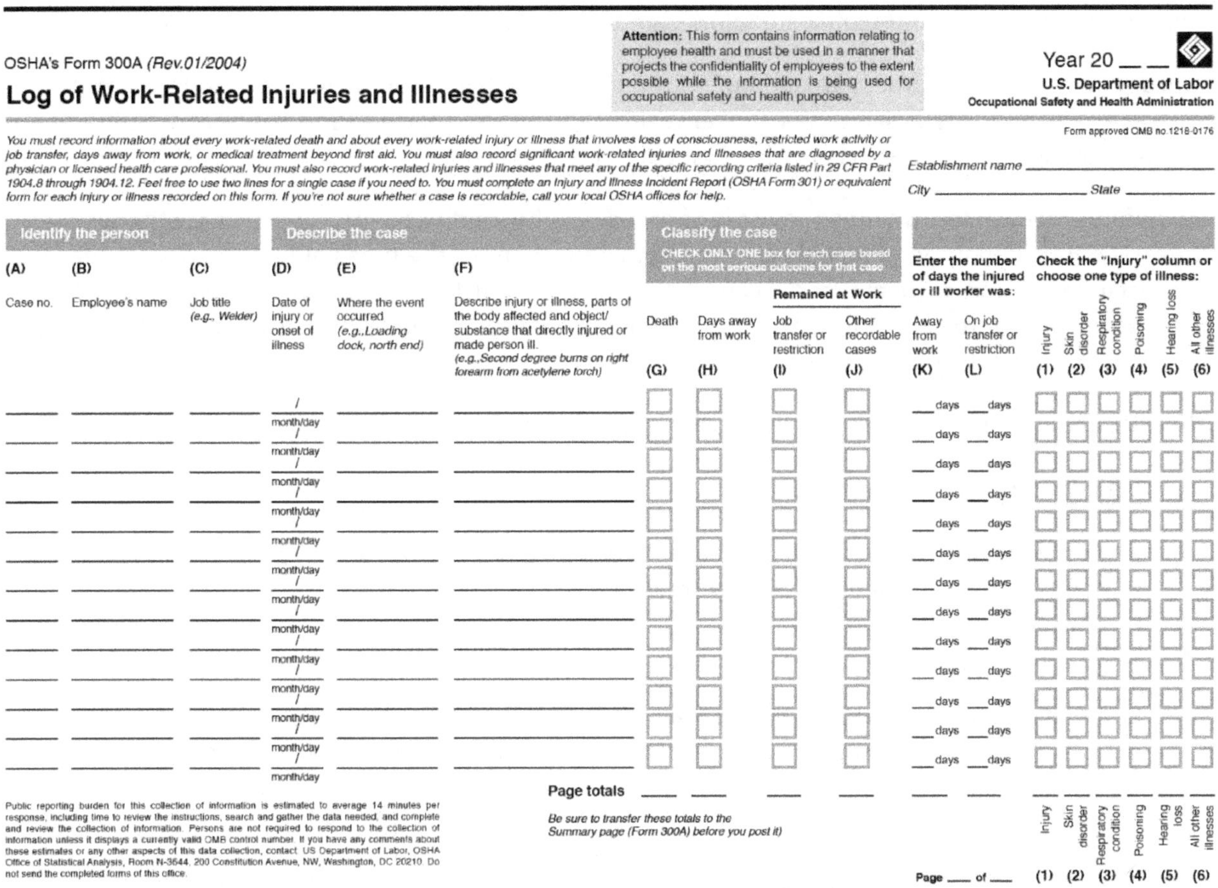

OSHA's Form 300A *(Rev.01/2004)*

Log of Work-Related Injuries and Illnesses

Attention: This form contains information relating to employee health and must be used in a manner that projects the confidentiality of employees to the extent possible while the information is being used for occupational safety and health purposes.

Year 20 __ __

U.S. Department of Labor

Occupational Safety and Health Administration

Form approved OMB no.1218-0176

You must record information about every work-related death and about every work-related injury or illness that involves loss of consciousness, restricted work activity or job transfer, days away from work, or medical treatment beyond first aid. You must also record significant work-related injuries and illnesses that are diagnosed by a physician or licensed health care professional. You must also record work-related injuries and illnesses that meet any of the specific recording criteria listed in 29 CFR Part 1904.8 through 1904.12. Feel free to use two lines for a single case if you need to. You must complete an Injury and Illness Incident Report (OSHA Form 301) or equivalent form for each injury or illness recorded on this form. If you're not sure whether a case is recordable, call your local OSHA offices for help.

Establishment name ______

City ______ *State* ______

Identify the person			Describe the case			Classify the case — CHECK ONLY ONE box for each case based on the most serious outcome for that case				Enter the number of days the injured or ill worker was:		Check the "Injury" column or choose one type of illness:					
								Remained at Work									
(A) Case no.	(B) Employee's name	(C) Job title *(e.g., Welder)*	(D) Date of injury or onset of illness	(E) Where the event occurred *(e.g., Loading dock, north end)*	(F) Describe injury or illness, parts of the body affected and object/substance that directly injured or made person ill. *(e.g., Second degree burns on right forearm from acetylene torch)*	(G) Death	(H) Days away from work	(I) Job transfer or restriction	(J) Other recordable cases	(K) Away from work	(L) On job transfer or restriction	(1) Injury	(2) Skin disorder	(3) Respiratory condition	(4) Poisoning	(5) Hearing loss	(6) All other illnesses
			month/day							__ days	__ days						
					Page totals	__	__	__	__	__	__	__	__	__	__	__	__
											Page __ of __	(1) Injury	(2) Skin disorder	(3) Respiratory condition	(4) Poisoning	(5) Hearing loss	(6) All other illnesses

Be sure to transfer these totals to the Summary page (Form 300A) before you post it)

Public reporting burden for this collection of information is estimated to average 14 minutes per response, including time to review the instructions, search and gather the data needed, and complete and review the collection of information. Persons are not required to respond to the collection of information unless it displays a currently valid OMB control number. If you have any comments about these estimates or any other aspects of this data collection, contact: US Department of Labor, OSHA Office of Statistical Analysis, Room N-3644, 200 Constitution Avenue, NW, Washington, DC 20210. Do not send the completed forms of this office.

Each incident case is then classified on the OSHA 300 log by the most serious event outcome for that particular case. There are four columns for recording the four classifications/outcomes in order of seriousness, the first of which is **death. Days away from work cases** are the most serious nonfatal outcome. The next most serious incidents are **job transfer or restriction cases** that involve modified duty.

For incidents resulting in missed workdays or modified-duty days, only days following the day of the incident itself factor into the classification. Assume that an employee is cut in the morning, receives stitches at noon, and then returns to work prior to the end of the workday, or perhaps even the next shift in the morning. This case will not be classified as involving days away from work because only days following the day of the incident are counted for this purpose; the same treatment applies to days with modified duty.

There will be times when an incident case involves both days away and modified-duty days. With these cases, it is important that the case be recorded as a single log entry and that it be classified as only one type. Since OSHA's recordkeeping scheme identifies days away from work as a more serious outcome than days with job transfer or restriction, such cases should be recorded only in the column for **days away from work cases**. If such a case is classified as both types, then at the end of the year when the column entries are added up, the total number of cases will be one more than what actually occurred.

Other recordable cases are the fourth and least serious event outcomes. These incidents do not result in any days away from work or modified-duty days but are still recordable because one or more severity thresholds have been met, for example, a loss of consciousness (even for just a minute), medical treatment, prescriptions, and significant injuries and illnesses.

Significant injuries include fractures and punctured eardrums. Significant illnesses include chronic irreversible diseases such as silicosis or cancer resulting from workplace exposures to contaminants—for example, prolonged exposure to concrete dust could cause silicosis.

OSHA defines medical treatment as any sort of procedure beyond basic first aid, such as stitches. However, OSHA's definition of first aid is not always equivalent to the types of first-aid treatments and topics discussed in a traditional first-aid class, and so a CHST must be familiar with OSHA's list of basic first aid because, if a treatment is not on the list, then it is deemed medical treatment by default and will require recording on the OSHA 300 log.

OSHA's list of basic first-aid treatments includes the following:

- Cleaning or flushing surface wounds
- Covering wounds with Band-Aids, gauze pads, etc. (wound closing devices such as sutures or medical glue are considered medical treatments)
- Wearing an eye patch
- Removing debris from the eye by flushing or with a simple cotton swab (using an eye magnet to extract a metal splinter and all other methods are considered medical treatment)
- Removing splinters from the skin by simple means, such as tweezers
- Draining fluid from a blister or relieving pressure from a toenail or fingernail by "drilling" with a heated needle or pin
- Using hot or cold packs
- Drinking fluids for the relief of heat illness
- General massages that are not directed by a health care professional (chiropractic treatments and physical therapy sessions are medical treatment)
- Wearing a finger guard
- Using nonrigid supports that do not immobilize a body part, such as elastic wraps or back belts (a rigid support that immobilizes a body part, such as a neck brace, is medical treatment)

- Tetanus shots (all other immunizations and inoculations administered in response to an incident are considered medical treatment)

- Using a nonprescription medication at nonprescription strength (prescriptions at prescription strength, even if available over the counter [OTC], filled or not, are medical treatment)

Treatments identified in the 1904 standard as first aid do not trigger OSHA recordkeeping themselves; however, an incident may still be recordable for another reason. For example, wearing a finger guard is considered first aid and not itself recordable, but if the finger is broken, then the incident must be recorded because a fracture is considered a significant injury, which is automatically recordable.

When days away from work or modified duty do not occur following the day of the incident, but a medical procedure takes place that is not identified as basic first aid, then the incident will be recorded in the **other recordable case** column. For example, an employee is cut in the morning, receives stitches by noon, and reports back to work the next morning would fall under this column.

In addition to first-aid events, diagnostic procedures also do not, themselves, trigger OSHA recordkeeping duties. Routine observation, counseling, x-rays, blood tests, and diagnostic medications administered following an incident are not considered medical treatments.

The purpose of diagnostics is to determine whether there is an abnormal condition that will require a particular course of medical action. The presence of an employee at a doctor's office is not a scenario that itself will trigger OSHA recordkeeping, though other factors, such as being prescribed painkillers at prescription strength, can still make the incident a recordable event.

An x-ray may be ordered to determine if there is a fracture. A fracture will result in an OSHA recordkeeping entry. However, if there is no fracture, then there will not be an OSHA recordkeeping entry unless some other recordable event takes place such as missing work for one or more days following the day of the incident, perhaps due to a high degree of pain and discomfort that instigated the x-ray in the first place.

At the end of the year, the total numbers of the four incident types that occurred will be entered on the OSHA Form 300A: the Summary of Work-Related Injuries and Illnesses. Establishment information, the average number of employees, and the total hours worked by all employees in the reporting year are also entered on the 300A. The 300A must be completed by February 1 of the next year and posted in the workplace from February 1 until April 30.

The establishment's Northern American Industrial Classification System (NAICS) code is also entered on the 300A. The NAICS code is a number of up to six digits that identifies the type of business activity that an establishment undertakes. All construction employers have an NAICS code beginning with 23.

OSHA Form 300A (Summary of Work-Related Injuries and Illnesses)

must be posted in the workplace from February 1 until April 30

OSHA's Form 300A *(Rev.01/2004)*

Summary of Work-Related Injuries and Illnessess

Year 20__ __

U.S. Department of Labor
Occupational Safety and Health Administration

Form approved OMB no.1218-0176

All establisments covered by Part 1904 must be completed by this Summary page, even if no work-related injuries or illnesses occurred during the year. Remember to review the Log to verify that the entries are complete and accurate before completing the summary.

Using the Log, court the individual entries you made for each category. Then write the totals below, making sure you've added the entries from every page of the Log. If you had no cases, write "0".

Employees, former employees, and their representatives have the right to review the OSHA Form 300 in it's entirely. They also have limited access to the OSHA Form 301 or its equivalent. See 29 CFR part 1904.35, in OSHA's recordkeeping rule, for further details on the access provisions for these forms.

Number of Cases

Total number of deaths	Total number of cases with days away from work	Total number of cases with job transfer or restriction	Total number of other recordable cases
______	______	______	______
(G)	(H)	(I)	(J)

Number of Days

Total number of days away from work	Total number of days of job transfer or restriction
______	______
(K)	(L)

Injuries and Illness Types

Total number of ______ (M)

(1) Injuries ______	(4) Poisonings ______
(2) Skin disorders ______	(5) Hearing loss ______
(3) Respiratory conditions ______	(6) All other illnesses ______

Post this Summary page from February 1 to April 30 of the year following the year covered by the form.

Public reporting burden for this collection of information is estimated to average 58 minutes per response, including time to review the instructions, search and gather the data needed, and complete and review the collection of information. Persons are not required to respond to the collection of information unless it displays a currently valid OMB control number. If you have any comments about these estimates or any other aspects of this data collection, contact: US Department of Labor, OSHA Office of Statistical Analysis, Room N-3644, 200 Constitution Avenue, NW, Washington, DC 20210. Do not send the completed forms of this office.

Establishment information

Your establishment name ______

Street ______

City ______ State ______ ZIP ______

Industry description *(e.g., Manufacture of motor truck trailers)*

______

Standard Industrial Classification (SIC), if known *(e.g., 3715)*

__ __ __ __

OR

North American Industrial Classification (NAICS), if known *(e.g., 336212)*

__ __ __ __ __ __

Employment information

If you don't have these figures, see the Worksheet on the back of this page to estimate.

Annual average number of employees ______

Total hours worked by all employees last year ______

Sign here

Knowingly falsifying this document may result in a fine.

I certify that I have examined this document and that to the best of my knowledge the entries are true, accurate, and complete.

______
Company executive — Title

( ) / /
Phone — Date

Health and Safety Programs

To protect the health and safety of employees, core elements make up the management philosophy and are the strength of programs in place. Since success starts at the top of an organization, effective safety and health programs should reflect this philosophy as well. Safety and health programs should be led and supported by management's commitment to leadership and accountability. Management develops written safety plans, sets organizational goals, and should set a good example for employees.

An effective safety and health program should consist, at a minimum, of the following primary elements:

- Management commitment
- Document control
- Employee involvement
- Hazard detection, assessment, and prevention with a hierarchy of controls
- Support relevant to contractors and subcontractors
- Worksite assessment and policy audits
- Continuous improvement with tracking and performance measures
- Daily and periodic inspection of machines and equipment
- Communication and training with accountability and enforcement

The application of these elements results in programs with clearly defined goals and reflects management philosophies with organizational accountability from top to bottom. Employee involvement and feedback is encouraged and helps to promote a positive safety culture. An "all-in" attitude serves to integrate safety and health goals into the organization's structure. Long-term tracking and solutions support and ratify continuous improvement efforts. Through collaborative efforts and imagination, cost-effective solutions can be presented that support an organization's bottom line.

Construction safety programs should consider health and medical aspects to meet the needs of the workforce. These include elements of nursing, first aid, disease control and prevention, and industrial hygiene. Health and safety programs should do the following to be successful:

- Maintain a healthy work environment
- Support disease diagnosis and immunizations
- Maintain an effective workers' compensation program that supports adequate healthcare and rehabilitation for work-related injury and illness
- Support open communication between employees' doctors and organizational health personnel
- Ensure periodic baseline testing and monitoring with respect to environmental exposures
- Provide drug and alcohol testing to protect personnel
- Provide health and fitness education and counseling
- Tailor health program elements to work processes
- Ensure workers' physical and mental capabilities are acceptable and fit for their job

Health and safety programs in construction must consider ongoing efforts to manage change. This adjustment process prevents new hazards from being introduced into the work environment. New hazards can stem from changes in equipment and technology, new design specifications, different materials used, changes to personnel affecting skill and experience, altered procedures, and modified standards and codes. An objective of change management is to ensure proper program maintenance with timely revisions and updates. Implementation and revisions must account for the following:

- Measures do not create an entirely new set of hazards.

- Measures do not increase the frequency, severity, and risk level of an existing hazard.
- Previously eliminated hazards are not reintroduced.
- Conceivable risk should be analyzed, and unforeseen/unexpected risk level should be minimized.

Applying Relevant Standards to Worksite Conditions

Many OSHA standards pertain to a particular construction activity; for example, 1926, subparts L, P, and S, address scaffolds, excavations, and underground work, respectively. OSHA expects that employees will be knowledgeable enough about construction health and safety standards to at least be able to find guidance on hazards that pertain to their work activity; there is no expectation that all employees, or even construction safety personnel, must be familiar with the entire breadth of OSHA's standards for the construction industry.

Personnel must only be familiar with OSHA standards within the scope of the construction activity with which they are involved. For example, a scaffold erector does not need to be familiar with excavation safety standards, and an excavating contractor does not need to be familiar with aerial lift safety. Additionally, an employee in an aerial lift bucket has no need to consult the standards for underground tunnel work.

However, a CHST should become acquainted with OSHA guidelines that pertain to the work activities of a subcontractor. While the general contractor may not conduct any excavation activities, his or her safety personnel will need to be sufficiently familiar with safety guidelines that pertain to a subcontractor's work activities so that hazards can be readily identified and eliminated.

General safety and health provisions that are common to all construction activity are described in 1926, subpart C. This section describes general requirements applicable to all construction employers, such as providing effective safety training, identifying and controlling hazards, maintaining safe and orderly construction sites by way of effective housekeeping efforts, and providing personal protective equipment (PPE). While this section describes a very small portion of the 1926 standards for the construction industry, a remarkable number of OSHA citations are issued for failing to meet these general requirements; a CHST must be very familiar with these general provisions regardless of the type of work being performed by the employer.

Inspections

Identifying and controlling hazards is a continuous process. When hazards are identified, they should be promptly and correctly eliminated. *Competent person* is a frequent term used in the construction industry, and OSHA calls for at least one competent person to be present on each construction site. Defined by OSHA, a competent person is "... one who is capable of identifying existing and predictable hazards in the surroundings or working conditions which are unsanitary, hazardous, or dangerous to employees, and who has authorization to take prompt corrective measures to eliminate them." OSHA's definition applies mostly to foremen, but any sufficiently capable employee can be deemed a competent person and authorized to identify and correct hazards. So long as hazard control is taking place, any number of competent persons may conduct inspections, whether they are a general laborer, foreman, or construction safety professional.

OSHA's guidance regarding inspections is to conduct "frequent and regular" inspections, allowing the contractor to determine the frequency and regularity of inspections on the job site. If hazards are corrected in a timely and reliable manner, the contractor's inspection regimen is working as intended. If hazards still exist after an inspection is conducted, the contractor's follow-up regimen has failed.

OSHA does not require inspections to be documented but mandates that identified hazards must be corrected. The absence of hazards proves that inspections and successful hazard-control practices are taking place.

Documenting inspection and hazard-control activities will provide valuable information. Even when only very minor hazards are identified that do not require work to stop in order to correct and may not even be mentioned at a safety meeting, documenting observations will provide safety administrators with data that they will use to analyze trends and allocate resources more effectively in the future.

The Board of Certified Safety Professionals believes that hazard control is most effective when the immediate supervisor is responsible for correcting identified hazards. The immediate supervisor is most likely to know the particular work area, its hazards, and the employees in that work area. Sometimes the supervisor is a CHST, but usually the CHST will work in an advisory role to assist managers to maintain maximum safety at all times. While a CHST may be capable of correcting the hazard, there can be unintended consequences. Employees may observe the CHST, a third party to their operation, as correcting hazards in their workplace and assume that this is just a corporate formality. They may surmise that if the hazard is not important enough for their supervisor to deal with directly, then why should they pay it any thought? Therefore, it is advisable that all project management be intimately involved with the identification, prioritization, and correction of all hazards.

Identifying Incident Trends

For application of its safety intervention strategies in the worksite, OSHA uses the injury and illness recordkeeping standard and written programs that must be produced by the employer should OSHA request them. The records the employer maintains are work-related fatalities, injuries, and illness, which include and identify hazardous industries, employers with poor safety programs, and trends for injury and illness. The CHST must realize how crucial proper recordkeeping is and how it validates a successful safety program. It also must be available for inspection during audits and is an excellent record of past hazards and their mitigation as applied to future activities. CAPA records can be included in this practice.

While small businesses (less than ten workers) remain free of routine illness and injury reporting, an increasing variety of commercial enterprises now fall under the routine and compulsory reporting for illness and injury required by OSHA. Previously discussed variances in worker ability to retain information must be identified, assessed, and remedied with the best possible means. Workplace injuries and illnesses remain notoriously underreported, perhaps because of workers' reticent to divulge personal information and possibly by the nature of the causal event. It is important that the workers perceive the CHST to be an impartial figure whom they may trust with personal and possibly embarrassing information. Illnesses, injuries, and resultant citations are directly proportional to the state of communication in the worksite. Citation categories include repeated, willful, serious, and other than serious; additional penalties may be assessed such as failure to abate, posting violations, or falsification of information.

Incident Rates

The data found on the OSHA 300A can be used to calculate the establishment's incident rates.

The "Days Away" rate is based on the number of cases involving days away from work following the day of the incident and is also referred to as DAFWII (Days Away From Work Injured and Ill), LWDI (Lost Workday Incidents), and LTIR (Lost Time Incident Rate).

The DART (Days Away, Restricted, and Transferred) rate is based on the number of cases involving days away from work following the day of the incident and also the cases involving days of restricted and transferred duty.

The TRC (Total Recordable Cases) rate is based on the number of all recordable cases in the calendar year and is also referred to as TRIR (Total Recordable Incident Rate).

Incident rates are used by many contractors as a factor in the qualification of subcontractors, and hiring clients and facilities also frequently review these rates as part of the bid review process when selecting contractors.

All three incident rates can be calculated with one simple equation: (*N* x 200,000)/EH

N represents the number of cases relevant to the particular incident rate desired. *EH* stands for "Employee Hours," the total hours worked by all employees in the reporting year. EH will be the same regardless of which incident rate is calculated. 200,000 is a fixed constant that never changes.

Assume a contractor has recorded one case involving days away from work, two cases with restricted or transferred duty, three "other recordable" cases, and 150,000 total employee work hours in the reporting year.

To determine the "Days Away" rate, the contractor will use 1 for *N*; the other values are fixed. To calculate the DART rate, the contractor will sum the total number of cases involving both days away from work with restricted and transferred duty. So, *N* will be 3 (1 plus 2). For the TRC rate, the contractor will sum "days away" cases, restricted/transferred cases, and also "other recordable cases." 1 plus 2 plus 3 equals 6, and that will be the value for *N* when calculating the TRC rate.

	(*N* x 200,000)/EH	**Incident Rate**
Days Away	$(1 \times 200{,}000)/150{,}000$	1.33
DART	$(3 \times 200{,}000)/150{,}000$	4.00
TRC	$(6 \times 200{,}000)/150{,}000$	8.00

Incident rates are calculated based on data entered on the 300 log and ultimately found on the 300A summary. For this reason, correct log entry is essential. If data on the 300 log is incorrect, then the incident rate will not be true; it may be correctly calculated based on the data provided, but the rate itself will be based on faulty data.

When tasked with OSHA recordkeeping duties, a CHST must understand that only incident events required to be entered on the OSHA 300 log should be entered. When incidents that do not meet the criteria for inclusion on the 300 log are entered, such as a case that involves nothing beyond basic a first-aid response (as defined by OSHA's 1904 standard), then the total number of recordable cases will be overreported, and the particular incident rate based on that data will be artificially high. This type of error has caused contractors to be disqualified from projects during the bid review process.

Evaluating Construction Means and Methods

Construction means and methods, or the prescribed, systematic way of accomplishing construction, are dependent upon the nature of the project and the manner in which the contractor intends to accomplish it. The contractor is responsible for employing the equipment and materials toward this end goal in the safest and most effective manner possible. The means and methods of construction include all of the day-to-day activity and equipment required to do the contracted work. It should be noted that means, methods, materials, and equipment are often outlined at the beginning of any construction project but may require frequent modifications throughout the construction process; therefore, all must be reviewed frequently for any necessary changes that need to be implemented. Frequently, construction projects involve "on the spot" modifications (for example, using different rigging equipment to accomplish a task than what was originally planned). All of these must be documented and changed appropriately in accordance to established rules and regulations so as not to cause added unintentional and unforeseen hazards.

The contractor generally chooses the means and methods he or she (or the industry in general) deems best suited to the project but in some cases, the project plan may call for alternate means and methods. In any case, special conditions and the specific scope of work may be called for in any given contract and must be reviewed prior to construction. It is the contractor's responsibility to ensure all subcontractors are aware of the overall means and methods outlined, along with any special considerations.

Project developers/engineers often do not apply safety considerations when planning a construction project. Consequently, a number of safety hazards are introduced early on during project design. As plans, documents, and designs approach finalization, such hazards run an ever-increasing risk of becoming inherent in the system, only to be discovered through triggered hazard events subsequent to the start of construction activities. This results in numerous negative impacts including injury and illness to workers, loss of assets, damages to the environment, delays from post-design retrofit processes, and the compounding costs associated with each. The effective provision of site safety requires that safety is included into project design. This includes materials and equipment selection, processes and procedures, and potential hazards stemming from process interferences of nearby and simultaneous work activities.

Safety in project design should include a baseline safety review for detecting and analyzing potential hazards. A baseline safety review should include both safety and design engineers to evaluate potential issues such as work and process hazards, industrial hygiene issues, fire safety, regulatory compliance, and emergency management. This step will provide safety and design engineers with critical information for proposing mitigation methods to enhance safety to workers and ultimately reduce costs. As many of the hazards encountered by construction workers are created by designers, by providing design engineers with this information, hazards that often result from their choice of materials and methods for the project can be significantly reduced.

A design and constructability review should occur when project design is approximately 90 percent complete. This evaluation should include an assessment of the ability for safe construction according to design and with respect to all parties involved. Representatives within this forum should include design engineers, safety engineers, estimators, schedulers, owners or upper management, and contractors and subcontractors. Information and feedback generated from participants can enable design and project managers to eliminate or mitigate hazards through additional plan and process modifications. Additional hazards or concerns that may remain can be monitored or communicated to relevant parties through contract language.

Leadership, Communication, and Training

Training Delivery Methods

Worksite training is integral to worker safety, and the CHST will use the available resources to ensure the worksite safety program is functional and consistently evolving. Worksite presence and intuitive interaction with workers by the CHST creates a dialogue with workers, while demonstrated empathy secures a connection. This connection manifests an increasing bond between worksite personnel. Once this bond is established, workers begin to stray from their comfort zones, which encourages honest dialogue and interpersonal conflict resolution, where individuals with disparate ideas begin to find common ground with one another. The desired end product is for people to connect on a positive emotional level.

In classroom settings, the CHST may be qualified to carry out certain training. If consultation or additional teaching resources are needed, the employer is obligated to provide them. The CHST should strive to further his or her certifications and training qualifications, as in-house trainings performed by an individual intimately aware of the worksite is preferred. It is equally important for the CHST to know the jobs that workers are performing in order to engage workers intelligently. Online resources are numerous, and the CHST will know and be able to reference these resources for training purposes. Assisting workers not versed in or overly familiar with online applications is the duty of the CHST.

OSHA and approved trainers provide online courses and training materials. On-site training includes videos, demonstrations, work observation, drills, and meetings. Off-site training may include classroom trainings, on- or off-site demonstrations, certification courses, and consultations.

No matter the diversity of languages, cultures, and educational levels on the worksite, employers must establish and maintain the programs necessary to achieve at least minimum-approved levels of operational safety that account for every worker on-site. In-house training may include worksite demonstrations, cross audits, presentations, drills, or safety meetings. Unique training needs may call for outside resources including consultants, specialists, and classroom courses.

Human Behavior Motivation Methods

An employer must make an honest and comprehensive assessment determining the individual mindset and the collective pulse of all project personnel. Behavior-based safety programs (BBSPs) will be key in the mitigation of assessed challenges and may include:

- A BBS that discerns the association between human behavior and injury prevention.
- A perception of how the attitudes and emotional state of workers influence their behavior in the workplace. It is the purpose of the BBSP in place to facilitate an avenue for constant exchange of official and informal information inclusive to all personnel.

Motivation is the incentive that drives people to behave in certain ways. To motivate employees to work toward the goals of the organization, leaders must understand what motivates people.

Motivational theories seek to explain what motivates people so that organizations can direct workers to behave in desired ways. Motivational theories can be content oriented or process oriented.

Content Oriented

Content-oriented theories seek to explain why people behave in certain ways.

Maslow's Hierarchy of Needs: As people fulfill their basic needs, they strive to fulfill greater needs higher up the hierarchy.

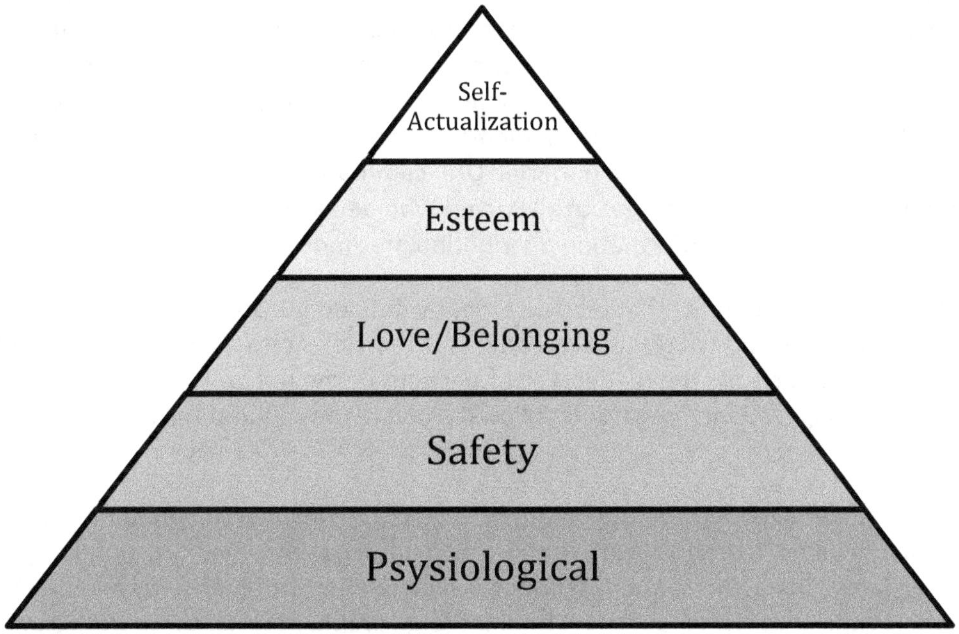

McClelland's Model of Realization: People are motivated by the need for relationships, power over their surroundings, or the ability to achieve goals.

Process Oriented

Process-oriented theories seek to understand how processes and psychological factors affect motivation.

Vroom Expectancy Theory: People are motivated by the value of the reward they will receive from achieving goals. These rewards can be internal (fulfillment) or external (promotion).

Goal-setting Theory: People are motivated when they set specific goals that are challenging to them and encouraged by the organization. Motivation is sustained through the feedback they receive as they work to achieve their goals.

Communication Strategies

Communication is the means by which all people receive and distribute information. Communication styles can be both verbal and nonverbal in nature. Most organizations rely on many forms of communication to accomplish their goals and objectives. Information distribution is accomplished through a vast array of communication strategies. With respect to purveying safety, focus must be placed on the use and effectiveness of three major communication strategies: *written*, *oral*, and *signage*.

It should be noted that many communication strategies depend on the inclusion of other strategic elements to be most effective. Examples are: written investigations and reports that require oral interviews; presentations and meetings that require written documentation and training aides for the audience; and signage that uses pictograms and symbols accompanied by succinct language.

Written Communication Strategies

Safety professionals are involved in a myriad of writing activities. The goal for each is to convey information in a manner as effective as possible for the prospective audience. The best communication strategy can vary depending on the situation. **Written communication** (e.g., emailing and texting) is often chosen for convenience and the speed at which messages can be sent to recipients. A notable advantage of written communication is that the content is more clearly defined than most verbal/oral conversations. With oral and signage strategies, elements can lack detail and be missed or forgotten since there is no opportunity to revisit the material. Oral communications can require additional time and coordination to disseminate messages to all parties who need to be informed. Following is a list of the mode and purpose of written information distribution methods:

- **Memos:** Written documents that include a clearly defined purpose and a summary stating the main intention. The body of discussion conveys the writer's message and may provide a short background with facts for the reader. Latter portions of the document should state a desired outcome with suggestion(s) regarding follow-up and recommendations. Memos should be one to two pages in length.

- **Lab and Field Reports:** Written documents designed to convince the reader of the validity of results and findings. A report's outline contains a purpose, introduction, and background section providing factual data. Subsequent sections include a description of steps and methods used. The report concludes with results and findings (e.g., graphs, charts, comparative analysis, etc.) and proposed recommendations. For the reader's convenience, an executive summary containing a condensed form of the major report aspects should precede the main report. An **executive summary** can be used when communicating large amounts of information to high-level associates. This is a brief summarization of important aspects from a larger document.

- **Incident/Accident Investigations:** Written documents to determine the primary or root cause of an incident and how to prevent recurrences. Additional purposes include fulfilling legal requirements, workers' compensation processing, assessing levels of compliance, and determining associated costs.

- **Meeting Minutes:** Modifiable documents that cover meeting agenda items and include names of participants and attendees, responsible personnel, notes pertaining to subject matter, date and time duration of meeting discussions, and action items discussed or addressed.

- **JSAs:** (See previous section covering JSAs.)

- **Surveys and Questionnaires:** A series of calculated questions (designed for yes/no answers or responses) applied to a rating or scoring system. Subsequent steps include data compilation and analysis to determine program and cultural successes and failures. Includes proposed recommendations to address deficiencies.

- **Operating Procedures and Instructions:** Written documents that provide step-by-step directions to learn an operation, complete a task, or perform an assembly. They consist of a combination of writing, pictures, labels, and symbols.

- **Data and Trend Analysis:** Gathering and analysis of data related to past events and behaviors over time for the prediction of future outcomes. This is achieved through tracking deviations, variances, and repetitious elements.

- **Emails:** Most notably, performance with writing emails is a matter of etiquette. Writers should keep subject matter to the interests of work and avoid criticizing or responding when angry. Writers should also be brief, formal, and proofread for errors and required edits. The writer should include a subject line appropriate to the content of the email.

Oral Communication

Oral communication is used to verbally transmit messages to an individual or group. Oral communications can be either formal or informal. Traditional types include face-to-face conversations or meetings, speeches, and telephone communications. Technology continues to offer new ways to transmit oral communications. Video communications such as Skype™ and podcasts are becoming more common in the workplace.

The most effective form of oral communication is considered face-to-face (in person) communication because it reduces the number of obstacles (also referred to as "noise") that impede a recipient's understanding of the message. This form allows for timely two-way communication versus the undesirable component of a delayed exchange in one-way written content. Face-to-face communication also allows for immediate feedback. Its flexibility in reiteration can result in improved understanding when compared to written communications.

With face-to-face communication, the deliverer of the message has the opportunity to assess the audience by observing reactions and has the ability to immediately confirm that the audience understands the message being delivered. In turn, the audience is able to ask questions and receive immediate feedback from the deliverer to clarify the message. By contrast, written communication does not afford the communicator the same opportunity to immediately reiterate or clarify the message being communicated. Situations usually exist where recipients may not be able to solicit the additional written information necessary for understanding the message. The following list includes the mode and purpose of oral information distribution methods:

- **Meetings:** Scheduled events to discuss and exchange information on a specific topic in a group setting. Meetings can be formal or informal.

- **Toolbox Talks:** Pre-task discussions on specific safety issues or the review of information before performing an altered task or undertaking a new process. Toolbox talks focus on one topic per discussion.

- **Three-Way Communications:** A process safety communication method where an actuating member verbally states his or her command of action to warn others. Other member(s) verbally respond in a similar way to confirm recognition. The actuator repeats the command before performing the action.

- **Presentations:** Provide training, guidance, or familiarity to a group on a particular topic.

- **Speeches:** Convey/Broadcast message(s) to a larger group.

- **Telephone Conversations:** Used to promptly provide information to an individual or group.

- **Interviews:** Performed to investigate or assess a person or incident.
- **Video Conference:** Used to share and exchange information with remote participants so they can see and hear other participants.

Signage and Symbols

Safety in construction activities incorporates a vast array of **signs and symbols**. These methods of communication are used to relay important cautionary information to both involved and uninvolved personnel. Hazard Communication Standards, lockout/tagout, crane and heavy equipment operations, and electrical safety are among the many processes where the use and application of these strategies are of vital importance.

- **Labels and Identifiers:** These elements are used to communicate elements such as ownership, manufacturer's criteria, expiration dates, potential hazards, or weight and load capacity.
- **Pictograms:** These are used as a visual communication method to caution or warn about various conditions. Pictograms are advantageous because of their capacity to defeat language barriers and achieve universal application.
- **Hand and Arm Signals:** These are physical movements/motions performed to provide assistance and guidance to help perform safe actions. Hand and arm signals are used to give commands, warn of impending hazards, increase alertness, and eliminate the interference of multiple processes.
- **Color:** Color is used to differentiate identity and function of otherwise similar components. Examples include the use of black, white, and green for wiring, and blue versus yellow to differentiate a water line from a gas line.
- **Alarms and Barriers:** These are used to warn, gain attention, and/or provide separation or isolation from an existing hazard. They employ device elements that impact multiple senses such as: sound and light amplitudes and frequencies; color; and placement with respect to lights and alarms, flags, lines, and cones.

Consulting with Equipment Manufacturers and Suppliers

The primary purpose of safety communication is to successfully convey information to the appropriate audience. Risk mitigation efforts and successes rely largely on the communication skills of the safety professional. Such skills are practiced through oral and written strategies and delivered via multiple modes and methods to all levels of an organization. With respect to safety culture, the core values of an organization are shared. However, differences in elements such as cultures, attitudes, types of work performed, and variances among departmental or individual objectives remain. This requires safety professionals to tailor safety communications specific to the entity (reader or audience).

The appropriateness of the selected communication mode, method, and delivery produces positive effects on attention, reception, comprehension, and retention of the message. As communication is critical for preplanning and design safety, the safety professional must also be able to receive and translate various types of documents.

These exist in the following forms:

- Contract documents containing safety information
- Design meeting and safety review documents
- Plans, drawings, signs, and symbols

To form effective communication strategies for speech and writing, the safety professional must refine various skills. For oral communication within forums such as meetings and reviews, the speaker should obtain knowledge of keywords for the situation. Effective speech may require practicing word choice and specific word usage to reveal strengths and weaknesses. The speaker should become skilled in the use of *tone* for words that need to garner special attention. This can include pausing for separation, reiterating words or phrases with emphasis to solidify the message, or saying words with an applied mood to ease tensions. The speaker should encourage the audience to reiterate key points and provide feedback for the given message to ensure comprehension.

Written communications must be tailored to the audience as well. The writer should avoid using terms unfit (too complex or oversimplified) for the situation. Written material used in conjunction with oral presentations should maintain consistency in word choice and avoid interchangeable words with similar meaning. In the event of post-design incidents, interviews conducted during investigations for corrective actions should express intent (why this is being done) and be recorded honestly and accurately to obtain a successful outcome.

Safety personnel should consider the audience when presenting data. In communicating site safety information, simpler explanations of data (percentages and pie charts) may be appropriate for one audience, while more complex explanations (line graphs and trend analyses) may be better suited for another. When presenting complex subject matter, the communicator may be required to clarify information orally, as opposed to overwhelming the audience with condensed written explanations. In these cases, the communicator must be wary of using distracting body language (accentuated movements/gestures) that may compromise the audience's mental concentration and focus on the complex data.

Manufacturers and Suppliers

Project leaders consult manufacturers and suppliers during the procurement stage of a project. This includes requesting bids, proposals, and information when planning what materials to purchase, and checking in regularly during the construction phase to ensure equipment and materials are received in a timely manner so the project remains on track.

Subject Matter Experts

Project leaders consult subject matter experts often during a project. During the planning stage, subject matter experts help project leaders to understand the true scope of the project, which human and material resources are required, which types of skills human resources should have, how much time should be allocated to perform the work, the budget required, and the risks common to that type of project.

During preconstruction, subject matter experts investigate work sites to perform industrial hygiene testing. During construction, subject matter experts check the site to ensure work is being performed safely and efficiently. Finally, after the project has been completed, subject matter experts might test the work to ensure it meets quality standards.

Confidentiality Requirements

In business, confidential information refers to protected information about the business that is not available to the public. Confidential information includes trade secrets, personal medical information, and personally identifiable information.

Trade secrets are defined under the Uniform Trade Secrets Act as information that holds economical value for a business due to being generally unknown to other businesses or the public. This includes proprietary information such as technical specifications documents, blueprints, formulas, product designs, algorithms, programs, techniques, processes, and information such as customer or supplier lists.

Trade secrets do not include other protectable confidential information such as employee agreements, non-disclosure agreements, or procurement documents.

Personal medical information is protected under HIPAA, which states that people have a fundamental right to privacy. Patient information must remain confidential. The patient must authorize any release of their medical information, and they may withdraw this authorization. Additionally, patients have the right to access their information at any time. Only the minimum amount of information required should be disclosed during legal record audits.

Personally identifiable information is information that can be used to identify, contact, or locate individuals. This includes financial, legal, employment, and education information. This information is protected under the law in some countries and by compliance requirements in others. Compliance rules often require that this information be encrypted, protected through data loss prevention tools, and stored according to rules that dictate how much information can be stored, where, and for how long. Finally, compliance rules regulate how much of this information can be shared within the organization and with third parties.

Security for both physical files and computer recordkeeping involves the control of access and storage. A project recordkeeping system collects, classifies, and stores a significant amount of private and intercompany information for employees, clients, contractors, and any exterior personnel employed on the worksite. Personnel given access to the recordkeeping database might be asked to sign a confidentiality agreement to access such.

BCSP Code of Ethics

The BCSP—along with the Council on Certification for Health, Environmental, and Safety Technologists (CCHEST) and American Board of Industrial Hygiene (ABIH)—are an alliance utilizing their collective expertise to increase the levels of education and proficiency in the workplace. These are some of the organizations that, in concert with OSHA, promote workplace professional development and pursuit of approved certifications. The CHST is encouraged to participate in the safety community and engage in higher learning opportunities, certification options, and the professional network the community provides.

The Board of Certified Safety Professionals (BCSP) is a peer certification board that sets standards for safety, health, and environmental professionals. Its core values are respect, accountability, leadership, and excellence.

The BCSP evaluates the technical competency, experience, performance, and educational credentials of professionals. It also administers exams to certify that professionals meet the standards defined in the certifications, and it continuously monitors performance to ensure certified professionals uphold those standards in the course of their work and their interactions with colleagues, clients, authorities, and the public.

The BCSP grants the following certifications:

- Construction Health and Safety Technician (CHST)
- Certified Environmental, Safety, and Health Trainer (CET)
- Certified Safety Professional (CSP)
- Associate Safety Professional (ASP)
- Occupational Hygiene and Safety Technician (OHST)
- Safety Trained Supervisor (STS)
- Safety Trained Supervisor Construction (STSC)
- Graduate Safety Practitioner (GSP)
- Safety Management Specialist (SMS)

The BCSP Code of Ethics ensures the professionals they certify adhere to a strict code of conduct to maintain the integrity and honor of the profession.

The Code of Ethics includes the following ideals:

- Hold paramount the safety and health of the people, the environment, and the property by informing employers, property owners, clients, the public, and the appropriate authorities in a timely manner if they discover dangers or unacceptable risks to safety and health.

- Be honest, fair, and impartial by balancing the interests of the client, property owner, employees, and the public and avoiding any practices that could deceive workers or the public.

- Issue public statements objectively and truthfully, and only when the information being presented is based upon strong evidence from competent subject matter experts.

- Accept assignments only when qualified to do so by experience and education in the specific field.

- Present only truthful and accurate information about previous experiences, education, and professional accomplishments to avoid misrepresenting or being deceptive about professional qualifications.

- Conduct relationships with transparency and professional integrity to avoid conflicts of interest that could lead to compromised judgement, and advise the BCSP of any misconduct by certified professionals.

- Act without bias toward age, gender, ethnicity, religion, nationality, sexual orientation, or disability.

- Provide service in civic affairs by sharing skills and knowledge and by taking action that promotes and improves the safety, health, and well-being of the community.

Developing Safety Training Requirements

As is the case for all industries, construction employers are expected to ensure their employees are trained to perform work activities in a safe and productive manner. OSHA holds employers accountable. No single training solution will meet the needs of every construction employer, and employers do not necessarily have to provide the training; they need to ensure that employees are sufficiently trained to perform their work tasks safely. Many opportunities to meet training needs are available: On-the-job training provided by the employer, union craft training, and courses at technical colleges are a few examples. Many employers use multiple sources to develop workers' skills.

Effective construction safety training results in improved, safe work performance. If training fails to improve performance, the training was a waste of the employer's resources. There are many reasons why training fails to result in improved safe work performance. For instance, the training content failed to meet the needs identified in the training-needs analysis and failed to assist the worker to develop the skills necessary to perform work safely. Also, the training may have been unnecessary. If a contractor is experiencing repeated eye and hand injuries because a foreman fails to enforce the contractor's rule to use PPE, more time spent in training is unlikely to make a difference. A formal needs analysis must be conducted to determine whether a training deficiency exists prior to recommending further employee safety training.

Employee safety training should always be documented. At a minimum, documentation should include the names of all trainees, date of the training, course title, course length, and instructor name and contact information. Ideally, the contractor should retain a biography of the instructor and a course description or outline for record-keeping purposes. Project management should actively consult applicable regulations regarding training and be aware of the minimum requirements in this regard.

However, training documentation alone does not ensure that employees are working safely; the best way to verify that training has resulted in improved safe work performance is to observe employees as they perform their tasks. Safe work performance is the only way to know whether training was effective, and a CHST must be able to communicate this to the managers they will be advising.

When OSHA evaluates the effectiveness of job safety training during an inspection, the evaluation is based on observations of work procedures and employee interviews; verification that training has been provided and training documentation itself play no role in this evaluation. For example, if employees are exposed to uncorrected fall hazards on a construction site, OSHA can justifiably assume the contractor failed to provide effective safety training. Violations for inadequate safety training are commonly cited by OSHA, and in many cases, documentation of training existed. As with all other types of citations, OSHA always cites the employer and not employees or supervisors.

As many employers focus on providing a safer work environment, deficiencies in safety programs and site-specific safety training for awareness and procedures remain. Employers must realize, even as all feasible risk controls are implemented and safeguards applied, the nature of construction work is inherently hazardous. Avoiding these hazards relies on effective safety training for employees.

Employers are required to provide training to their employees as outlined in OSHA 1926.21, Safety Training and Education. Because processes are numerous and unique to each project, employers must develop their own specific means and methods to meet these requirements. Critical aspects of effective site safety include employers' knowledge and understanding of process-applicable regulations as well as management's commitment to safety and a well-developed safety culture within the organization. For the development of site-specific training, employers must practice the following:

- Acquire input from knowledgeable trainers, industry professionals, and regulatory sources.
- Ensure training elements are appropriate to the target audience.
- Provide adequate training tools, suitable facilities, and safe conditions/methods.
- Prioritize training needs based on employees who are subject to the highest hazard risk.
- Ensure employees are active in demonstrating skills and work practices and providing feedback.
- Develop plans for scheduling and duration, level of employment (in years and experience), and design procedures for new hires and retraining as necessary.
- Deliver training (as part of preplanning efforts) prior to the start of work.
- Conduct job hazard analysis for high-risk processes on a job site.

Training is critical for the safe performance of a vast array of construction processes. Preplanning or in situ training requires employers, management, and safety personnel to remain up to date with current standards, methods, and best practices. Other important aspects for ensuring site safety include pre-task toolbox talks, scheduled meetings, and reviews to reveal strengths and weaknesses in order to promote continual improvement, and equal and consistent enforcement of violations.

Maintaining Applicable Documentation

Documenting safety information and processes is a critical part of safety. OSHA, other governing agencies, as well as insurance providers, require documentation. Documentation is essential for control, tracking, and maintenance of various safety processes and will be requested in the event of an external (OSHA) post-incident inspection or safety audit. A multitude of safety processes and supporting elements are required by law to be written and documented. Such items include confined-space procedures, emergency action plans, respiratory protection, and hazard risk assessment. OSHA states that organizations should consider the following elements for achieving regulatory compliance:

- Knowledge of standards applicable to an organization's work site and program needs
- Documentation for certification of hazard assessment including the particular site or facility, hazards, controls, procedures, and training
- Continual maintenance, review, and updates for training records
- Completion and recording of OSHA-mandated certifications and assessments

For developing documentation, an organization should first prepare a safety mission statement that demonstrates commitment to protecting the health of their employees. Effective document development relies on teamwork from a safety committee comprising key departmental personnel that direct the collection and sensible organization of required document elements. This committee should also develop approaches for meeting/exceeding regulatory compliance and for reaching the goals and objectives of company policies.

Part of document maintenance and control includes tracking and identifying recent revisions to the document. Organizations must have a system for identifying which written plans and documents are most current. A document coordinator should be selected to control and maintain documentation. Using outdated documents results in lost time, wasted efforts, and potentially harmful impacts. Organizations should maintain documents in a centralized and secure location (e.g., electronic repository) and make them available to employees and associates as necessary.

Crane Certifications

These consist of four major required certifications: load rate markings, load rate test reports, rope inspections, and hook and chain inspections.

- **Load rate markings:** This stipulates that the manufacturer designed and approved the load rating clearly marked on the side of the crane; for multiple hoist cranes, the load rating must be clearly shown on each individual hoist.
- **Load rate test reports:** These are available for viewing and specify load rate tests carried out for modifications or hook repairs. They also allow crane operators to know the current crane capacity for loads.
- **Rope inspections:** In addition to the requirement that employers must hold all documentation and certification for crane rope inspections on file, there is a mandated monthly rope inspection, and an inspection is required if the rope has been out of service (OOS) for a month or more.
- **Hook and chain inspections:** These require daily and monthly inspections documenting the date of inspection, the inspector's signature, and an identifying marker (serial number or alternate identification) on the hook or hoist chain.

Training Records

OSHA mandates that employers ensure proper safety training (communication, information, or instruction) for all worksite personnel; all training must be documented, and this documentation must be signed by all participating employees. In addition, these training records must be kept on file for a minimum of three years (OSHA).

Training documentation will be reflected in safety meetings and toolbox sign-in sheets, the safety training matrix, drills, and other manners of safety performance tracking and worker certifications. Training documentation will be preserved in written and digital form with access provided to select personnel who will monitor, update, and track the documentation with reference to future training needs. It is important to note that training must also be part of any ongoing CAPA records.

JHA

JHA remains a primary tool for worksite hazard identification, assessment, and mitigation. The CHST must be involved with all steps of the JHA process: Participate in JHA meeting, sign off on JHA sheet, ensure appropriate copies are disbursed to concerned personnel and jobsite, and close out the JHA once the task is completed. It is necessary to keep a well-organized file with all past and present JHAs, and it is essential that all personnel sign off on these and all other safety training documents.

Injury Logs

OSHA utilizes injury and illness recordkeeping to not only try to track workplace injuries and illnesses, but also to track the manner in which these issues are addressed and resolved by the employer. OSHA uses these logs to formulate and plan their audits, shape their enforcement program, and improve regulatory standards. OSHA considers injuries and illness to be one of the more poorly and underreported aspects of worksite operational safety.

CAPA Records

CAPA records should be kept on file in accordance with all international, federal, and state regulatory bodies. Such records are subject to audit by those bodies. Each must be in direct response to a hazard or an incident that has occurred and must identify the hazard or the incident and be preventive (pre-incidental) or corrective (post-incidental) in nature. Each must identify the personnel responsible for the action and include the steps taken to prevent or correct the identified problems. Each CAPA record must be managed as a project itself and include deadlines for the implemented actions and project management sign-off is mandated upon completion.

Determining Training Requirements

Worksite hazard identification training activities on the worksite are deemed by NIOSH to be effective forms of positive intervention. These programs, coupled with OSHA voluntary training guidelines and incorporating site-specific guidelines, will help to maintain a dynamic safety program.

OSHA, as dictated by the Labor Act of 1970, will establish and supervise worksite safety programs to train workers and employers in the recognition, avoidance, and prevention of accidents. An employer must institute a worksite safety training program and site-specific safety plan to ensure each employee is compliant and/or certificated with the specific requirements of their duties or trade (accommodations must be made for language, cultural, and education levels, as every employee must receive the minimum approved training required to perform their task). Any and all training can, and should, be considered a vital part of corrective/preventive actions and then fully documented in associated CAPA records as a step toward resolution.

To determine training requirements for work site personnel with varying levels of skill, education, experience, and language proficiency, leaders conduct the following assessments:

- Organizational
- Individual
- Training Needs

The organizational assessment examines the organization to determine the current level of performance of the organization to identify areas in which performance can be improved. The assessment also measures which skills, education, and knowledge are required in order for the organization to achieve the desired level of performance. This assessment helps identify strengths and weaknesses in the organization, so leaders can enhance their strengths and work to eliminate or weaknesses.

The individual assessment examines the performance of the workforce. Leaders assess groups or individual workers in the areas in which performance requires the most improvement. The assessment identifies which groups or individuals require the most resources and which specific skills they are lacking. During this assessment, leaders may determine all the workers in a particular group are lacking specific skills, or that only one or two require additional training.

The organizational and individual assessments identify the specific skills and knowledge the workforce requires to perform at the desired level, and which groups or individuals require the most training. The training needs analysis determines how to fill the gaps in knowledge.

Once the training needs have been identified, leaders must identify the best way to deliver training based on the limitations of the workforce. Workers who lack prerequisite skills might require additional training before they can train with the rest of the workforce. In this case, leaders might determine that the workers can gain the knowledge they need from computer-based training or a one-day seminar. Some workers might require training in different languages, or they might be spread out in different geolocations. Leaders might decide to create computer-based training courses recorded in multiple languages that are accessible anywhere in the world.

At-Risk Conditions and Behaviors

Left unchecked, task-related risk behaviors lead to the majority of worksite incidents. These behaviors have resulted in the worksite safety culture. The CHST must be capable of correctly identifying worksite at-risk behaviors and consulting any resources deemed necessary to make a sound decision on how to proceed if the situation warrants follow-up or investigation. The potential delicacy of some workplace situations calls for the exercise of sound judgment by the CHST to ensure a fair and unbiased observation.

Task-related risk behaviors can be defined as those workplace habits that directly contribute to hazards and incidents. Thorough review of past workplace incidents through the hazard identification, assessment, and mitigation process described above will help the CHST determine how to proceed in the current environment. Any and all safety-related behavior critical to the success of the project must be identified and communicated across the workforce. Workgroup safety should be stressed as part of the on-site culture. The workforce should feel as if communication regarding at-risk behaviors and conditions is acceptable to relate to management and vice versa. Project management should also regularly review system- and process-related barriers, along with physical or chemical barriers, to on-site at-risk behaviors and conditions. This will help increase overall visibility to potential incidents and help project management control.

The goal of identifying at-risk behaviors and conditions is to minimize incidents, reduce injury rates, reduce workers' compensation costs, and positively contribute to a workforce that values safety. Failing to identify at-risk behaviors and conditions will result in critical, and potentially fatal, losses.

As behavior-based safety (BBS) is increasingly integrated into the work environment, the CHST must possess sound knowledge of the existing worksite risk reduction programs in place and how they correspond to the BBS Program (BBSP). BBS is not a stand-alone system, but a tool designed to tie in to the worksite safety programs in place. As the connection between behavior and action is borne out in the worksite, BBS is increasingly recognized as integral to the HSEMS (HSE Management System). Ninety percent of incidents recognize that the behavior of personnel directly involved was the causal agent, whereas the remaining ten percent were influenced by behavior of personnel indirectly involved. The BBSP will require engagement by all personnel with the CHST helping to facilitate such interaction. A BBSP enables a group of workers to identify, assess, and amend worksite behavior in order to reduce behavior-related risks.

For this program to work, the CHST and all safety personnel are responsible for nurturing a no-blame atmosphere at the worksite (Dupont's STOP™ safety observation cards are a good example of a no-name, no-blame worksite hazard identification tool). It is essential the CHST maintain an impartial approach when monitoring worksite behavior, and auditing and investigating incidents.

Antecedent Behavior Consequence (ABC) is the behavior model on which the BBSP is based. This theory concludes that for every behavior, there are one or more factors that cause it and one or more factors that persuade or dissuade the repetition of it.

As observation is a prevalent facet of BBS, which must be carried out in a nonintrusive and neutral manner. On the worksite, workers and management come together and organize work observations and/or departmental cross audits (a JHA remains an excellent tool for daily behavior observation and cross-department interaction but is contingent on management participation and support). This can be accomplished with worker engagement that breaks the ice and renders the typical impediments such as embarrassment, pride, and ego from the process. Encouraging management and workers to perform work observations or cross audits in concert also has the added benefit of encouraging the human interaction needed to create empathy between people who do not work side by side.

The following terms may be helpful to know when identifying at-risk behaviors and conditions:

- **BBS:** This studies the effects of worksite environment, circumstance, and situation on a worker's behavior and regard for safety.
- **At-risk:** A potentially dangerous condition or action.
- **Antecedent:** The causative factor of an action.
- **Behavior:** The manner in which a person acts.
- **Consequence:** The result of an action.

The responsibility of the CHST is to observe, perceive, identify, and mitigate aberrant worksite behavior before it instigates a hazard. This is done through earning the trust and respect of workers on the construction site, which allows the CHST to perceive the pulse of the worksite and be the beneficiary of timelier, voluntary information.

Recognizing Imminent Danger Situations

Imminent danger is a situation in which death or serious injury could be reasonably expected to occur. This includes situations in which workers might lose limbs or the ability to use body parts, lose consciousness, lose a large amount of blood, or experience severe heat stroke. This might also include situations in which a worker comes into contact with a deadly disease or experiences serious skin damage from a toxic substance.

Imminent dangers include fires or explosions, chemical or hazardous material spills, collapse of building or structures, outbreaks of deadly diseases, release of deadly toxins into the air or water supply, workplace violence, natural disasters, and medical emergencies.

Some situations that might present an imminent danger include the following:

- Employees mishandling hazardous chemical or biological material
- Employees using heat-producing tools in the vicinity of flammable gases or liquids
- Employees working on the side of a busy road without proper traffic control
- Employees working on elevated platforms or above moving machinery
- Employees working in extreme heat or cold without proper clothing
- Employees handling toxic substances without appropriate protective equipment
- Employees operating tall machinery or equipment near active powerlines
- Employees using power tools with frayed or damaged cords
- Employees using power tools in the vicinity of water
- Employees working in trenches in which a high probability of a cave-in exists
- Employees exposed to asbestos, lead, silica, hexavalent chromium, and other toxic compounds

In order to recognize situations that present an imminent danger, leaders should inspect work sites and work spaces. In the inspection, they identify potential safety hazards that could result in imminent danger, so they can eliminate unsafe conditions and prevent accidents through safety precautions. Precautions might include removing obstacles around elevated surfaces, eliminating flammable material from the area, ensuring proper shut-off and start-up procedures are in place, providing proper protective equipment, ensuring workers handle chemicals according to SDS, inspecting work spaces for toxic compounds, inspecting power tools for damage, and posting signs to warn of specific dangers in the area.

If a hazard results in an imminent danger despite precautions, leaders should have procedures in place to stop work and respond to emergencies.

Coaching Personnel

The CHST and all construction project personnel must have skills in a variety of areas in order to run an efficient and well-maintained worksite. These skills include, but are not limited to, the following:

- Coaching
- Communication
- Application of best practices
- Conflict resolution
- Hazard identification/prioritization/mitigation in a timely manner
- Implementation of corrective and preventive actions (CAPAs)

All project personnel must be able to effectively coach construction project workers as to safe behaviors. As mentioned previously, workplace behavior is the most common cause of hazards and subsequent injuries. Appropriate review of past incidents will help project management members establish a baseline from which to begin coaching efforts. All project management personnel and the CHST must have up-to-date knowledge regarding safe behavior in the workplace and conduct both group and individual training/coaching sessions as necessary.

Communication flow is important. Both verbal and written communication are vital in the workplace. All workflows and related job directives must be documented on file in accordance with regulatory bodies. All such pertinent information must be verbally disseminated as well. The workforce must feel free to report any potential hazards and make incident reports as they occur in accordance with established policies and procedures. Also important in successful communication and leadership is appropriate conflict resolution. Conflicts are likely to arise over time between any two people. The ability to respectfully and positively address and manage the conflict can improve teamwork and the bond between team members.

Project management personnel and the CHST must be aware of the industry's best practices and how they are applicable to the project at hand. This information can, and should be, disseminated both in writing and during frequent training review sessions. Best practices may include any job-related tasks associated with physical and environmental tasks.

It is critical that all workplace personnel can rank, prioritize, and mitigate identified hazards in a timely manner. A workflow applicable to this process should be drafted, documented, and kept on file at all times. Once again, personnel should be trained in these process steps and can implement them should the occasion arise.

Any and all hazards or incidents must have CAPAs associated with them. These CAPAs should carry a CAPA plan. The CAPA plan is best limited to one hazard per plan. It should identify the hazard, the root cause of any related incidents as the result of the hazard, the personnel responsible for affecting a change, what that corrective change is, how it will be implemented, by when the corrective action will be implemented (a deadline), and what preventive actions will be put in place moving forward so the hazard will be mitigated or eliminated altogether.

Accessing Relevant Current Information

Keeping up-to-date with change is vital for maintaining compliance in programs and policies. Many organizations often overlook aspect of change management and continual improvement efforts. Implementing safety elements based on standard and code modifications demonstrates commitment and due diligence. Conversely, failing to maintain and apply current knowledge results in employers with deficient programs, thereby prescribing outdated work processes, static training methods, and associated best practices. In the event of an OSHA inspection, or a program audit due to a serious incident, such practices can result in serious consequences for the employer where negligence can be a factor.

Examples of recent changes are OHSA's widely known **Hazard Communication Standard (HCS)** (1910.1200 – 1926.59) and the **final ruling on crystalline silica** exposures. Such changes impact a multitude of elements which are applicable to in-place employee programs, processes, and training methods. With respect to the silica ruling, the following are program elements subject to review and revision:

- Access control plan/regulated areas
- New engineering and administrative controls
- Respiratory protection/associated action level and permissible exposure limits
- Medical surveillance
- Hazard communication training
- Personal protective equipment

The safety professional is responsible for keeping up with OSHA standards and best practices. Hazard impacts, new developments in technology, improved methodologies, and scientific research are all part of elements that necessitate rule modifications and updates. Knowledge of the appropriate tools and resources can simplify this overwhelming process. A complete understanding of modifications, applications, exceptions, and other factors can be gained by viewing OSHA interpretations.

In today's technological environment, computers and the Internet are by far the most effective tools for accessing current information. Their main advantage over any paper publication is that updated information can be accessed at light speed. However, the sheer volume of Internet material coupled with deficiencies in searching skill can limit these advantages. Obtaining current information requires skills in the use of search terms, methodologies, and features provided within the website. Safety professionals and management can use the following sources for keeping up to date with ongoing changes in standards, codes, and best practices:

- OSHA.gov
- Safety.BLR.com
- J.J. Keller Online
- SafetyDailyAdvisor.com, available at ehsdailyadvisor.blr.com
- Other sites developed for state-run compliance

Among the tools listed above, OSHA.gov is the most complete tool a skilled safety practitioner can use to access current information. OSHA's website provides features and options including a comprehensive home page, advanced search, and a full A to Z index of topics. Data & Statistics offers searching by establishment, inspection data, and frequently cited standards by **Standard Industrial Classification (SIC)** code or industry type. Users can also obtain data related to accident investigations and fatality statistics.

Practice Questions

1. What resource is required to be in place prior to the start of any construction project?
   a. Materials storage
   b. Acquisition of scaffolding
   c. Prompt medical attention in case of serious injury
   d. All power tools

2. With respect to contract safety, which statement is NOT true of the host employer?
   a. The host employer is responsible for communicating general site conditions to the contractor.
   b. The host employer must hire the contractor with the best safety record.
   c. The host employer must evaluate the contractor's safety performance history.
   d. The host employer should ensure safety responsibilities are shared.

3. Which statement is true regarding site owners, contractors, and subcontractors (all employers)?
   a. All parties should focus on site-wide hazards and corrective actions.
   b. Any party can contract away the responsibilities protecting employees from hazardous work conditions to another party.
   c. All parties are responsible for ensuring clarity and understanding of contract language.
   d. Best practice for all employers involved in the work includes assuming the highest level of safety responsibilities for an entire multi-employer work site.

4. Which of the answer choices best completes the following sentence?
A crane must be certified on the date of ____________ by the government to be structurally and mechanically capable of performing in the manner specified by the manufacturer.
   a. Deployment
   b. Mobilization
   c. Certification
   d. Manufacture

5. Although safety responsibilities are shared on multi-employer work sites, which party is generally responsible for practicing safety?
   a. The contractor
   b. The host employer
   c. The subcontractor
   d. The safety department

6. When a loss control auditor carries out a worksite audit, which of the following documents will need to be available?
   I. Site-specific HSE plan
   II. Records of crew safety training
   III. Sign-in sheets for tailgate meetings
   IV. Crane inspection records

   a. III
   b. II, IV
   c. I
   d. I, II, III, IV

7. Which of the following comprises prequalification, pre-task planning, risk assessment, orientation, monitoring, and post-work evaluation?
    a. Baseline safety reviews
    b. Project life cycle
    c. Contract bid process flow
    d. A complete project manual outline

8. Which of the following provides safety-related input and feedback from diverse key personnel, broadens the reach of safety, and increases safety preplanning effectiveness for better safety and health outcomes?
    a. Management commitment
    b. A good workplace safety culture
    c. Communication policies
    d. Multidisciplinary teams

9. Which of the following requires a full assessment as part of preplanning and is an often overlooked and underestimated facet for safe work practices and sharing of safety responsibilities?
    a. Baseline safety reviews
    b. Toolbox talks
    c. Safety performance reviews
    d. Safety culture assessments

10. Which of the following is the maximum PEL for worker lead exposure?
    a. Fifty micrograms of lead per cubic meter of air during an eight-hour period
    b. Five hundred micrograms of lead per cubic meter of air during an eight-hour period
    c. Twenty-five micrograms of lead per cubic meter of air during an eight-hour period
    d. Ten micrograms of lead per cubic meter of air during an eight-hour period

11. In an effort to reduce work-related musculoskeletal disorders at the worksite, which of the following ergonomic mitigation factors can be utilized?
        I. Redesign tools
        II. Modify equipment
        III. Use alternate materials
        IV. Update work processes

    a. I, IV
    b. II
    c. III
    d. I, II, III, IV

12. Fitting a worker with an approved respirator is an example of which type of control measure?
    a. Engineering
    b. Substitution
    c. Administrative
    d. PPE

13. Subcontractors should practice safety and responsibility in which of the following manners?
    a. One that is focused on their own knowledge and expertise of safety processes.
    b. One that is relevant to their scope of work as stated in a contract.
    c. One that assists employers and contractors in hazard control within all project locations.
    d. One that concedes all subcontractor safety responsibilities to the employer and the main contractor.

14. Which does NOT apply to lifting?
    a. Keep the load within ten inches of your body.
    b. Pivot the load at the base of the spinal column.
    c. Keep your back straight, and lift with your legs.
    d. Use a solid, two-handed grip.

15. Which two controls place the focus of risk mitigation in the following order: (1) on the worker (and worker behaviors), and (2) on the source of the hazard, respectively?
    a. (1) Redesign, (2) administrative
    b. (1) Engineering, (2) personal protective equipment
    c. (1) Substitution, (2) redesign
    d. (1) Administrative, (2) redesign

16. Which safety process method involves a written model for pre-task planning that detects and addresses all possible hazards related to a hazardous task where multiple job steps are performed?
    a. Risk assessment matrix
    b. Job safety analysis
    c. Hierarchy of controls
    d. Abilities assessment form

17. Which of the following hazard assessment methods use frequency and severity to determine the level of hazard risk? Resultant risk levels can be qualitative or quantitative.
    a. Failure mode and effect analysis
    b. "Why?" methodology
    c. Risk matrices
    d. Job safety analysis

18. Which type of control includes worker training or the selection of specific personnel for a task?
    a. Substitution
    b. Engineering
    c. Administrative
    d. Personal protective equipment

19. What does PEL mean?
    a. Potential electrical limit
    b. Permissible exposure limit
    c. Protected electrical line
    d. Platform elevation level

20. Inspections of machines, material, and equipment should be led and monitored by which of the following?
    a. The safety professional
    b. Upper management
    c. Qualified engineers
    d. The maintenance department

21. Model organizations that are part of OSHA's Voluntary Protection Program can be researched/referenced to do which of the following?
    I. Enhance existing programs and policies
    II. Improve workplace safety culture
    III. Receive preferential treatment during audits
    IV. Promote safe work practices

    a. I and II only
    b. I, II, and III
    c. I, II, and IV
    d. I, II, III, and IV

22. Hazard risk level related to many industrial hygiene exposures is related to which of the following parameters?
    a. Frequency and severity
    b. Cause and effect
    c. Concentration and duration
    d. Physical and chemical

23. Which of the following does NOT apply to the physical and chemical properties of chemicals?
    a. Flammability
    b. Autoignition temperature
    c. Coagulation
    d. Toxicity

24. To which limits must an organization achieve legal compliance with regard to hazardous substance and chemical exposures?
    a. Threshold limit values (TLVs)
    b. Permissible exposure limits (PELs)
    c. Recommended exposure limits (RELs)
    d. Only PELs and the TLVs that are more stringent than PELs

25. Which choice below fits the respective roles of (1) detecting potential hazards, and (2) communicating safety roles and responsibilities?
    a. (1) Workers, (2) safety professionals
    b. (1) Management, (2) workers
    c. (1) Workers, (2) management
    d. (1) Safety professionals, (2) management

26. Which of the following can result from safety program incentives in the form of cash awards?
    I. Lower incident rates
    II. Increased awareness
    III. Nonreporting of incidents

    a. I
    b. II
    c. I and II
    d. I, II, and III

27. Which is the most likely result of larger companies and more work disciplines on a construction site?
    a. Hazards of simultaneous operations
    b. Improved safety records
    c. Better safety cultures
    d. Higher quality of work

28. Which reactive process requires the correct utilization of a series of "Why?" questions?
    a. "What-if" analysis
    b. Root cause analysis
    c. Failure mode and effect analysis
    d. Safety culture analysis

29. Injury risk is a combination of likelihood and which of the following?
    a. Time of day
    b. Consequence
    c. Worksite obstacles
    d. Inclement weather

30. Follow-up procedures for corrective actions include which of the following?
    a. Root cause analysis
    b. "What-if" analysis
    c. "Why?" analysis
    d. Tracking

31. At what point of project design completion should a design and constructability review occur?
    a. Approximately 50 percent
    b. Approximately 75 percent
    c. Approximately 90 percent
    d. 100 percent

32. Which analysis method is conducted to answer the question, "Are/were the results of the corrective action applied worth the time, money, and effort of implementation?"
    a. "What-if" analysis
    b. 50 percent design and constructability review
    c. Cost-benefit analysis
    d. Root cause analysis

33. Which of the following illustrates engineering controls?
    a. Worksite HSE signage
    b. Video safety training
    c. Equipment barriers
    d. Posted safety notices

34. Which of the following does NOT reflect routine compulsory records maintained by the employer?
    a. Work-related illness
    b. Work-related injury
    c. Work-related fatality
    d. Work-related confrontations

35. A chain of command is selected for reacting to emergency situations. These individuals should be selected based upon which of the following?
    a. Years of experience with the organization
    b. Level of title, i.e., beginning with upper management
    c. An individual's capacity to perform under stress
    d. Job responsibilities

36. Which of the following is most critical for maintaining effectiveness and employee respect for safety programs and policies?
    a. A safety committee comprising key personnel
    b. Management's knowledge of people and processes
    c. A good workplace safety culture
    d. Enforcement

37. For evaluating general site conditions and relevant information, which group among the following is most critical for building design and construction?
    a. Topographical, hydrologic
    b. Soil, topographical
    c. Vegetation, soil
    d. Hydrologic, soil

38. According to the OSHA requirements for stair construction, the stair slope or angle must fall between which of the following?
    a. 20 and 40 degrees
    b. 30 and 40 degrees
    c. 30 and 50 degrees
    d. 40 and 50 degrees

39. A ladder is extended a total of 50 feet in length and positioned to an upper landing. Which of the following are the requirements for this extension ladder?
    a. The ladder must extend 3 feet above the upper landing with a minimum overlap of 3 feet.
    b. The ladder must extend 4 feet above the upper landing with a minimum overlap of 3 feet.
    c. The ladder must extend 3 feet above the upper landing with a minimum overlap of 4 feet.
    d. None of the above

40. Which of the following is NOT associated with fall protection equipment?

I. Harnesses and lanyards
II. Restraints and guardrails
III. D-Rings and anchor points

a. I
b. II
c. III
d. I and III

41. With regard to workforce diversity and a multicultural public, which form of communication would be most effective in communicating a jobsite hazard?

a. Written
b. Oral
c. Signage
d. Email

42. Which of the following is true regarding silica-associated illnesses and hearing damage?

a. It needs to be treated within twenty-four hours of contraction.
b. It is associated with ingestion-based illness.
c. It can be prevented but not repaired.
d. It can be delayed with proper medical therapy.

43. A trench is designed with a slope in excess of 45 degrees. The soil in this trench was tested and has a compressive strength greater than 1.5 tsf. Which statement best describes this situation?

a. The trench is in stable rock.
b. The soil in this trench is most likely Type B and dry.
c. The trench cannot be deeper than 10 feet without being designed by a registered engineer.
d. The soil in this trench could be either Type A or stable rock.

44. A confined space has the potential to entrap or entangle a worker. Because of this condition, what must the supervisor do in the confined space before work activities begin?

a. Acquire a permit.
b. Ensure that the entrant can't be trapped or entangled.
c. Ensure that the tripod/harness rescue equipment is in place.
d. Practice the work to be performed in an unconfined space.

45. Which demonstrates the most effective concept for applying a *hierarchy of controls*?

a. Consider personal protective equipment (PPE) requirements for hazards in a new process, and then develop affordable methods of substitution, engineering, and administrative controls.
b. Assess and eliminate all conceivable hazards in the design stage. Then, continue down the hierarchy, ending with PPE to eliminate all remaining hazards and residual risk.
c. Assess and eliminate all conceivable hazards in the design stage. Then, assess and minimize the remaining risk to acceptable levels, applying the hierarchy while considering prioritization and feasibility.
d. Apply the combination of substitution and engineering controls. Then, apply administrative controls and regular employee training as fail-safe redundancies.

46. Of the following, which is NOT defined as a primary physical hazard of a construction site?
    a. Falling accidents
    b. Electrocutions
    c. Concussions
    d. Struck-by accidents

47. Which of the following is NOT included in the hierarchy of controls?
    a. Administrative
    b. Operative
    c. Engineering
    d. Substitution

48. Hard hats are vital for protecting against electrical shocks, bumps, and impacts. OSHA standards have adopted the testing and rating criteria of which entity?
    a. American Society of Testing and Materials (ASTM)
    b. National Institute of Occupational Safety and Health (NIOSH)
    c. American National Standards Institute (ANSI)
    d. Underwriters Laboratories (UL)

49. Which process adjustment does NOT align with the philosophy of addressing ergonomic hazards?
    a. Using an adjustable scaffold to change work and/or material height as needed.
    b. Selecting the right worker with skills, strengths, and experience for a particular job.
    c. Equipping a delivery truck with a hydraulic lift gate.
    d. Fitting job elements to meet the needs of the worker.

50. For applying controls in risk mitigation efforts, which statement best reflects PPE considerations?
    a. All higher control options must be exhausted before employers consider PPE.
    b. To set a good example, PPE can be donned by employees on a voluntary basis.
    c. PPE is a last resort for risk elimination.
    d. PPE selection depends on the process.

51. A sound level meter detects 87 dBA within several construction site areas. These levels exceed any previous readings by 3 dBA. Decibel levels are consistent throughout the course of the normal workday, and earplugs are available. In this situation, the employer will reach compliance through which of the following actions?

I. Applying effective controls or developing a *Hearing Conservation Program (HCP)* due to exceeding the action level.
II. Achieving noise attenuation efforts of 3 NRR, to reduce ALL exposures to 84 dBA (below action level), thereby eliminating HCP requirements.
III. Providing hearing protection due to overexposure.

    a. I
    b. II
    c. III
    d. II and III

52. Control measures used to lower exposure to silica in the workplace may include which of the following?

I. Ventilation
II. Restrictions for time on-site
III. Facial masks with OSHA-approved fine mesh screens
IV. Approved respiration gear

a. II, III, IV
b. I, II, IV
c. II, III
d. I, IV

53. As part of fall prevention systems, guardrails must be able to withstand how many pounds of force?

a. 100 lbs.
b. 200 lbs.
c. 500 lbs.
d. 5000 lbs.

54. Which form of written communication should omit fine details, but provide data and results, while not exceeding two pages in length?

a. A memo
b. An email
c. An executive summary
d. A lab/field report

55. Which form of written communication should contain a clearly defined purpose, provide background and facts for the reader, state intentions, and finish with a desired outcome, while not exceeding two pages in length?

a. A memo
b. An email
c. An executive summary
d. A lab/field report

56. If an employee wears a dust-filtering face piece as voluntary self-protection from exposure, the employer should note this as which of the following?

a. An indication of employee involvement in a positive safety culture
b. A violation of company policy
c. A trigger point to develop a written respiratory program
d. A good example for others to follow

57. A compounding factor of a trench collapse can include which of the following?

a. The use of timber instead of aluminum in shoring methods
b. Dense soils at greater depths
c. Not using a qualified engineer with excavations deeper than 10 feet
d. Increased depth and the presence of water

58. For a confined space to require a permit, what conditions must be present?
   a. The space must be large enough that a person can fully enter to perform work.
   b. The space must not be intended for permanent or continuous occupancy.
   c. Means of entry and exit must not be limited.
   d. None of the above

59. Which hazard mitigation method is the most proactive?
   a. Failure Mode Effect Analysis
   b. Heinrich's Domino Theory
   c. Chain of Events Theory
   d. Fault Tree Analysis

60. Of the following items, which is NOT used to quantify risk-assessed hazards?
   a. Severity
   b. Consequence
   c. Impact on key performance indicators
   d. Likelihood of occurrence

61. Which most closely represents the correct sequential flow for a general *lockout/tagout (LOTO)* procedure?
   a. Notify and shutdown → isolate energy → verify
   b. Identify energy sources → isolate energy → shutdown → verify
   c. Apply LOTO devices → shutdown → isolate residual energy
   d. Relieve stored energy → verify effectiveness → apply LOTO devices

62. Which of the following is NOT an example of an administrative control?
   a. Job rotation
   b. Monitoring
   c. Fence installation
   d. Using a less hazardous substance

63. A professional registered engineer must design protection systems if an excavation is greater than how many feet in depth?
   a. 10 feet
   b. 15 feet
   c. 20 feet
   d. 25 feet

64. What is the most important component for an effective safety program?
   a. Continuous improvement efforts
   b. Responsibility, accountability, and consistent enforcement
   c. Employee involvement
   d. Management commitment

65. Which of the following is performed to determine the root cause of an incident, prevent recurrence, and fulfill legal requirements?
   a. Job Safety Analysis
   b. Safety Audit
   c. Cost/Benefit Analysis
   d. Incident Investigation

66. In a workspace atmosphere, a monitoring device's alarm is set to trigger if oxygen levels fall outside what range of percentages?
    a. 21, 78
    b. 12, 18
    c. 10.5, 21.5
    d. 19.5, 23.5

67. Which of the following is NOT mandatory for a worksite crane?
    a. Rope inspection
    b. Hook and chain marking
    c. Load rate marking
    d. Air conditioning in operators' cabin

68. Which safety communication strategy is exclusive to verbal exchanges among all parties involved?
    a. Hand and arm signals
    b. Pictograms
    c. Three-way communication
    d. Presentations

69. Which of the following is true regarding scaffolding?
    a. Tie-ins are required with a unit that is 30 feet tall and 6 feet wide.
    b. Working and walking areas must be fully planked, at least 12 inches wide.
    c. Mudsills are not required when erected on solid, level, and dry asphalt.
    d. All construction projects employ scaffolding at some stage of the work.

70. Accidents occur when hazards and which of the following are combined?
    a. Number of workers
    b. Exposure
    c. Work from heights
    d. Restricted areas

71. Before a worksite hazard is assessed, which of the following must occur first regarding the hazard?
    a. It must be mitigated.
    b. It must be categorized.
    c. It must be identified.
    d. It must be assigned.

72. As part of egress from trench excavations deeper than four feet, ladders must be located so that employees have no more than how many feet of lateral travel?
    a. 20 feet
    b. 25 feet
    c. 35 feet
    d. 50 feet

73. For hazards of confined spaces, which of the following presents the greatest concern?
    a. Hazardous atmospheres
    b. Lockout/Tagout and electrical hazards
    c. Entrapment and entanglement
    d. Blunt force trauma

74. A safety practitioner is assisting in developing controls to address hazards of falls. Several employees are simultaneously at risk. This is a new job, so any system can be used, and elements of feasibility are not an issue. Which system would be most effective for addressing the hazards?

a. Multiple retractable systems consisting of state-of-the-art technology
b. Horizontal lifelines with safety nets as a redundancy
c. Restraints and warning lines in conjunction with a safety monitor
d. Compliant guardrails

75. Approximately what percentage of incidents is the result of human behavior indirectly associated with the task?

a. 10 percent
b. 90 percent
c. 20 percent
d. 80 percent

76. With regard to the hierarchy, what do lower level controls and higher level controls address respectively?

I. PPE, materials
II. Behaviors, equipment
III. Duration of exposures, safety through design

a. II
b. III
c. II and III
d. I, II, and III

77. Which safety process/document has both proactive uses and reactive uses, can be used for training purposes, and shows the breakdown of a larger task into smaller elements with associated hazards and corrective actions?

a. Incident Investigation
b. Field Report
c. Job Safety Analysis
d. Data and Trend Analysis

78. Which information-gathering process is used to determine the alignment of goals and attitudes, and to assess the overall safety culture of a workforce from top to bottom?

a. Job Safety Analysis
b. Meeting Minutes
c. Survey/Questionnaire
d. Data and Trend Analysis

79. With regard to confined space activities, which statement about the attendant is NOT true?

a. The attendant can enter the space only when relieved by another qualified attendant.
b. The attendant must review the feasibility of rescue equipment.
c. The attendant must never perform any other activity that interferes with those relevant to confined space duties.
d. The supervisor can assume the role of attendant.

80. OSHA requires proper housekeeping on all construction sites in order to maintain a safe and orderly workplace. Where can specific guidelines for housekeeping be consulted in the OSHA standards?
   a. 1910, subpart D, Walking-Working Surfaces
   b. 1910, subpart E, Means of Egress
   c. 1926, subpart C, General Safety and Health Provisions
   d. 1926, subpart M, Fall Protection

81. Which is the most correct statement regarding effective construction safety training?
   a. Employee safety training must be documented in order to meet OSHA requirements.
   b. The best safety training course is usually the one with the most descriptive outline.
   c. Employer-provided safety training is always the best training option for new employees.
   d. Improved safe work performance is more important than formal documentation retention.

82. Which of the answer choices best completes the following sentence?
Along with crane certifications and logbooks, ________________ must be available for audits and inspections.
   a. Operator manuals
   b. Paint codes for hook and chain
   c. Certified load limit indicators
   d. Lift rope specifications

83. Which is the best advice a CHST can provide to a construction manager?
   a. Recommend that supervisors be held accountable for correcting hazards.
   b. Suggest using the OSHA Form 301 when documenting hazards.
   c. Clarify that inspections must be carried out at least once per week.
   d. Encourage that a work stoppage be mandatory for all identified hazards.

84. Which of the following is NOT a purpose of a JHA or JSA?
   a. To aid in the identification of task-specific hazards
   b. To provide workers with SDS information relevant to the task
   c. To resolve post-task workers' compensation issues
   d. To receive worker feedback pertaining to the anticipated task

85. Which of the following worksite mandates is NOT exercised by OSHA?
   a. Conducting unannounced worksite inspections
   b. The OSHA area manager working with an employer to institute voluntary safety initiatives and programs
   c. Reviewing off-site worker safety practices
   d. Requesting documentary recordkeeping of worksite safety programs

86. What is the minimum amount of electrical current that can be considered fatal?
   a. 7 milliamps
   b. 70 milliamps
   c. 7 amps
   d. 70 amps

87. What is the best practice with electrical cords?
    a. Run cords through wall holes instead of along floors to eliminate tripping hazards.
    b. Fasten cords to walls with staples instead of nails to avoid damaging insulation.
    c. Protect cords from accidental damage when running them through doorways.
    d. Use only electrical tape when making effective repairs to damaged insulation.

88. Which of the following statements is correct?
    a. Only a competent person may approve the removal of a ground prong from an electrical cord.
    b. A continuous path to ground is NOT necessary for powered tools that are GFCI-protected.
    c. Equipment manufactured prior to 1971 is exempt from OSHA's electrical safety standards.
    d. The use of tools that are NOT grounded is permissible so long as they are double-insulated.

89. Which of the following is NOT a form of asbestos?
    a. Crocidolite
    b. Chrysotile
    c. Amosite
    d. Vermiculite

90. Which statement regarding Ground-Fault Current Interruption (GFCI) devices is true?
    a. A difference in current between the hot and neutral return wires will cause the GFCI to trip.
    b. The interrupt device will actuate when 70 mA of current or more are detected going to ground.
    c. GFCI is a component of an assured equipment grounding conductor program in construction.
    d. A power cord must have a ground prong (third wire) for GFCI to function as intended.

91. Which of the following is a correct statement?
    a. OSHA's safety requirements for aerial lifts can be found in 1926, subpart M, Fall Protection.
    b. Fall protection is required on scaffolds when six feet or more above the next lower working surface.
    c. Fall protection on aerial lifts is always required, even when less than six feet above the ground.
    d. Stairways and ladders do not require fall protection because OSHA's general duty clause does not apply.

92. Which of the following is a fall restraint system?
    a. Standard guardrail
    b. Safety net
    c. Shock-absorbing lanyard
    d. Full-body harness

93. What is true regarding a personal fall arrest system (PFAS)?
    a. Body belts require that their anchorage supports at least 5000 pounds per attached worker.
    b. The maximum total fall distance allowed in a personal fall arrest system is six feet.
    c. A PFAS is preferred to fall restraint because the components undergo more reliability testing.
    d. Only components approved by OSHA's product safety division are permitted in fall arrest systems.

94. What should always be done with floor holes found on working surfaces in construction?
    a. Place standard guardrails on all open sides of the hole.
    b. Notify employees of the means selected to eliminate the hazard.
    c. Set up flagging lines four feet from the hole in all directions.
    d. Install a floor hole cover at least 72 pounds per square foot in strength.

95. Which of the following is a true statement regarding fall protection on scaffolds in construction?
   a. Fall protection must be provided when six or more feet above the ground or the next lower working surface.
   b. Scaffold cross bracing alone may be used in place of a complete scaffold guardrail system in construction.
   c. Employees working below a scaffold do NOT need to wear hard hats when scaffold toe boards are used.
   d. The top rail in a scaffold guardrail system must be installed between 38 and 45 inches from the deck.

96. Which of the following factors is NOT associated with a successful behavior-based safety program?
   a. Improved worker communications
   b. Improved participation in safety culture
   c. Direct contact between workers and OSHA
   d. Increased worker information feedback

97. Which of the following behavior-based safety definitions is used to describe the causative factor of an action?
   a. At-risk
   b. Consequence
   c. Behavior
   d. Antecedent

98. Which is the safest guidance a CHST can recommend for work involving aerial lifts?
   a. Ensure that guardrails used for fall protection are compliant with 1926 construction safety standards.
   b. Require that all employees in aerial lifts wear a personal fall restraint system when working over water.
   c. Even when body belts are permissible for fall restraint, use a more expensive fall arrest system instead.
   d. Lanyards should be six feet in length, since OSHA allows a maximum fall arrest distance of six feet.

99. Which of the following is a correct statement?
   a. During inspections, OSHA's evaluation of safety training is NOT based on training documentation.
   b. OSHA allows no more than one person to be designated as a competent person at a construction site.
   c. Only licensed electricians and apprentices are allowed by OSHA to repair electric power cords.
   d. A personal fall arrest system (PFAS) is always required when working on a scissor lift platform.

100. Of the following, which is NOT a site inspection employed by OSHA?
   a. Imminent danger
   b. Fatality/catastrophe
   c. Supplemental inspection
   d. Follow-up

101. During a routine inspection, a CHST identifies a minor safety hazard that has repeated itself a few times over the past month. Which is the best course of action?
   a. Exercise Stop Work Authority, convene a mandatory site-wide safety meeting, and address the hazard.
   b. Ask the foreman to correct the hazard, document the event, and report the situation to management.
   c. Identify the employee who created the hazard, address the hazard, and proceed with disciplinary action.
   d. Correct the hazard, document the hazard, and then address it at the next safety committee meeting.

102. How frequently does OSHA require first-aid kits in construction to be inspected?
   a. Every shift
   b. Daily
   c. Weekly
   d. Monthly

103. What is meant by "Universal Precautions?"
   a. Sending all personnel on the construction site to receive basic first-aid training, including CPR and AED
   b. Requiring a hepatitis B vaccination for all employees who will be designated first-aid providers
   c. Developing an exposure control plan for all first-aid and medical personnel at a construction site
   d. Treating all bodily fluids as containing blood-borne pathogens or other potentially infectious materials

104. Which of the following is the most hazardous form of external radiation to a worker?
   a. Beta particles
   b. Alpha particles
   c. Gamma rays
   d. Microwave ovens

105. What is true of emergency planning?
   a. Notifying emergency management personnel is always the first action taken in an emergency.
   b. Employer emergency action plans (EAPs) are addressed by OSHA's HAZWOPER standard.
   c. The first priority in an emergency is to safeguard personnel, equipment, and processes.
   d. Emergency alarms in construction must have a sustained sound level of at least 105 A-weighted decibels.

106. A CHST sees the letter B on a fire extinguisher. Which class of fire does this letter represent?
   a. Flammable Liquids and Gases
   b. Electrical Equipment
   c. Ordinary Combustibles
   d. Combustible Metals

107. Which is true of fire extinguishers in an outdoor construction environment?
   a. OSHA requires fire extinguishers on construction sites to be inspected at least once per week.
   b. *Dry chemical* and *multipurpose dry chemical* can be said to refer to the same thing.
   c. An extinguisher must be located within 10 feet of five or more gallons of flammable materials.
   d. There must be one fire extinguisher for every ten employees assigned to a construction site.

108. Which topics does OSHA require to be addressed in blood-borne pathogens training?
  a. Basic first aid, cardiopulmonary resuscitation (CPR), and automated external defibrillator (AED)
  b. All relevant provisions of 1926, subpart D, Occupational Health and Environmental Controls
  c. Housekeeping, personal protective equipment (PPE), control methods, and hepatitis B vaccine
  d. All topics approved by the American Red Cross, Bureau of Mines, or an equivalent curriculum

109. Which of the following is a correct statement?
  a. To facilitate evacuation, an assembly point should always be fixed for a project's duration.
  b. An emergency action plan (EAP) is NOT required for employers with ten or fewer employees.
  c. OSHA requires the hepatitis B vaccine for all personnel designated as first-aid responders.
  d. At least one fire extinguisher must be adjacent to a stairway in a multistory construction site.

110. What is the objective of an incident investigation?
  a. Arrive safely and care for the wounded.
  b. Reduce the contractor's liability obligations.
  c. Identify regulatory and legal violations.
  d. Prevent a similar incident from occurring.

111. Which of the following is NOT associated with a safety training matrix?
  a. Safety performance tracking
  b. Worksite standard operating procedures
  c. Hazard mitigation tracking
  d. Personnel assigned to hazard mitigation and closeout

112. Which is the best guidance a CHST can provide in regard to interviewing a witness following an incident?
  a. The witness must understand the disciplinary focus of the interview.
  b. The incident site itself is generally the best place to conduct an interview.
  c. Interviews should take place no sooner than three days following an incident.
  d. The interview should ideally take place in the safety administrator's office.

113. Which statement regarding incident review processes in construction is true?
  a. When beginning an incident investigation, the first priority is to arrive safely and size up the situation.
  b. A safety administrator is the lead person in most routine incident investigations in construction.
  c. There is typically one causal factor in an incident; the occurrence of multiple causal factors is rare.
  d. Meetings with union representatives must take place before an incident investigation begins.

114. Which of the following is a correct statement?
  a. Employees should NOT sign chain-of-custody documents in order to maintain anonymity.
  b. General contractors and subcontractors should develop plans independently to avoid confusion.
  c. The number of employees who will handle investigation evidence should always be minimized.
  d. A CHST should assume that senior management is well informed during safety presentations.

115. Which is true of the OSHA Form 301?
  a. It must be completed by no later than the end of the day of the incident.
  b. It is the employer's Log of Work-Related Injuries and Illnesses.
  c. It must be completed within 7 calendar days of the day of the incident.
  d. It is the employer's Summary of Work-Related Injuries and Illnesses.

116. Which event requires an incident to be entered on the OSHA 300 log?

I. The rabies vaccine is administered following a rat bite that took place.
II. Blood is withdrawn to determine whether an employee contracted hepatitis C.
III. A tetanus shot is administered following a cut caused by a rusty nail.

a. I
b. II
c. III
d. I, II, and III

117. Which of the following best identifies the benefits of daily safety walks performed by the CHST?
a. Networking with contractors in view of future work opportunities, accomplished by seeking them out during safety walks
b. Engaging in a meaningful dialogue for CHST and workers to find common ground on safety issues
c. Choosing the proper side in case of an incident so as to favor the side aligned with management
d. Patrolling the worksite and documenting worker mistakes in the presence of others to set an example

118. Which of the following is a correct statement regarding OSHA recordkeeping duties?
a. OSHA requires that all construction employers record injuries and illnesses.
b. Only incidents involving days away from work or modified duty are recordable.
c. OSHA's 1904 standard identifies draining fluid from a blister as medical treatment.
d. General contractors do NOT record the injuries of subcontractor employees.

Please refer to the following data for questions 119 and 120:

Year	Cases with Days Away from Work	Cases with Days Away, Restricted, and Transferred	Total Recordable Cases	Employee Hours
2014	4	4	12	480,000
2015	5	8	14	560,000
2016	4	7	14	510,000

(N x 200,000)/EH

119. Which statement is correct?
a. The DART rate increased for each year in which data was recorded.
b. The TRC rate was the same for 2014 and 2015 and increased in 2016.
c. Both the DART and TRC incident rates were at their highest in 2014.
d. The DART rate was lower than the "Days Away" rate in every year.

120. What was the DART rate in 2016?
a. 1.57
b. 2.75
c. 11.13
d. 17.85

121. Which is the best OSHA recordkeeping advice that a CHST can provide?
   a. Make an effort to complete the OSHA 301 incident report the same day as the incident.
   b. Record nonrecordable events on the OSHA 300 so more data is available for review.
   c. Only incidents involving medical treatment should be entered on the OSHA 301.
   d. Prohibit non-safety personnel from performing OSHA recordkeeping duties to ensure privacy.

122. Which of the following is a correct statement?
   a. An administrator's office is a good venue for interviewing incident witnesses because it is so neutral.
   b. OSHA requires that every identified hazard be documented so that it can be properly corrected.
   c. A scaffold must be secured when its height is more than four times greater than its width at the base.
   d. Class D fire extinguishers are desirable in construction because they can handle every class of fire.

123. Where can OSHA's recordkeeping standard be located?
   a. 29 CFR 1915
   b. 29 CFR 1910
   c. 29 CFR 1926
   d. 29 CFR 1904

124. Which of the following does NOT apply to a struck-by event?
   a. Struck by a rolling object
   b. Struck by a flying object
   c. Struck by a fellow worker
   d. Struck by a swinging object

125. Sparks can fly up to how many feet?
   a. 45
   b. 50
   c. 35
   d. 40

126. Which of the following types of hazards occurs when a worker strains their back due to lifting heavy materials?
   a. Struck-by
   b. Physical
   c. Health
   d. Stress

127. Which of the following is true of powder-actuated tools?
   a. They can go off and shoot nails if the safety is disengaged.
   b. They operate like a loaded gun.
   c. The types of powder-actuated tools used in the U.S. fire at over 428 ft./s.
   d. They can be fired from a distance of up to 5 ft.

128. Which of the following effects of extreme heat causes permanent tissue damage?
   a. Heat rash
   b. Heat stroke
   c. Both of these
   d. Neither of these

129. Under which of the following types of hazards does the GHS classify electric shock?
  a. Health hazard
  b. Physical hazard
  c. Environmental hazard
  d. Imminent danger

130. Which of the following key concepts of the incident command system refers to limiting duties based on the role of an agency?
  a. Modular organization
  b. Management by module
  c. Unit of control
  d. Span of control

131. Which of the following does builder's risk cover?
  a. Water damage due to flood
  b. Accidents on the job
  c. Structural damage due to high winds
  d. Claims against the contractor for faulty workmanship

132. Which of the following is classified as a trade secret?
  a. Suppliers lists
  b. Employee addresses
  c. Medical records
  d. Procurement documents

133. How does the BCSP Code of Ethics require professionals to avoid conflicts of interest?
  a. Avoid practices that could deceive workers or the public.
  b. Balance the interests of the client, property owner, employees, and public.
  c. Conduct relationships with professional integrity.
  d. Present only truthful information about their qualifications.

134. Which of the following assessments is used to identify areas of strength and weakness in the performance of the workforce?
  a. Training needs analysis
  b. Workforce assessment
  c. Organizational assessment
  d. Individual assessment

Answer Explanations

1. C: The items to be in place prior to project implementation are based on the pre-project risk assessment and are fundamental to the safety of the work to be done. Catering, tools, and scaffolding are operational and material concerns. These can be changed before, during, or after a project.

2. B: The host employer must hire the contractor with the best safety record. The host employer is responsible for ensuring that safety is a joint effort among all parties and communicating general site conditions to contractors and subcontractors along with evaluating contractors' safety performance history. Although host employers are required to vet contractors, they are not bound to select a contractor based upon best safety performance results.

3. C: All parties have the responsibility of ensuring clarity and understanding of contract language. Subcontractors should not be involved in the corrective actions process. The subcontractor should only report potential hazards to the contractor or site owner. OSHA 1926, subparts A and B, mandates that all employers must protect their employees from hazardous work conditions. Best practice includes the host employer focusing on monitoring safety and ensuring safety is shared among entities. The contractor should not absorb the entirety of site safety, and the subcontractor should focus on safety responsibilities exclusive to the scope of the work.

4. D: Mobilization and deployment are characteristic of project-related crane activities. Date of manufacture certification deems a crane is fit to be used on the worksite by OSHA.

5. A: Although safety responsibilities are shared, the main contractor is primarily responsible for practicing safety, as the contractor normally performs the bulk of the work processes. Many exceptions can alter responsibilities, but such items must be included within the contract and be agreed upon by all parties.

6. D: Not all listed documents will necessarily be requested, although they all need to be available to auditors. The project site-specific safety plan is the primary document and declaration for the assessment and mitigation of known worksite hazards. The auditor can determine currency and type of training for workers with review of training documentation. The toolbox meeting and JSA sign-off sheets will be requested should an incident or audit occur. Once again, for purposes of audits or incident reporting, crane records for load testing, maintenance, and certification are to be kept and available.

7. B: Project life cycle is described by a chronological flow of contract phases and work/safety processes from conception to finished work. Safety reviews and bids are part of contract development but exist as individual elements. Project manuals, if present, contain the whole of all contract documents related to the project.

8. D: Multidisciplinary teams focus on how safety relates to multiple departments and their respective processes. Management commitment is critical for leadership and supporting a multitude of program elements, processes, and policies. A good safety culture is helpful for the respect of safety and the performance of safe practices. Communication is vital for relaying information. Although Choices *A, B,* and *C* are all important, they are not specific to safety across multiple departments.

9. D: An organization's knowledge of how employees view safety will provide the organization with indicators of perceived worker behaviors before work activities begin. This allows an organization to make any necessary modifications to management practices and/or program elements. A toolbox talk is a pre-task communication promoting hazard awareness. Safety and performance reviews reveal data but cannot predict worker behaviors and shared responsibilities for safety.

10. A: Fifty micrograms of lead per cubic meter of air during an eight-hour period defines the OSHA-prescribed exposure limit; the remaining answers are wrong.

11. D: All answers entail applicable methods designed to provide safeguards against work-related injuries. The redesign of tools counteracts the forces of torque, weight, and impact on the worker and improves tool usability. Equipment modifications for the worksite may include noise reduction or installment of guards to prevent worker injury. Alternating materials are concurrent with attempts to replace materials containing lead, silica, asbestos, and other harmful chemicals with less harmful ones. The manner in which a job is physically done can invariably be improved. Every task should be reviewed and evolved to better suit the ergonomic and physical environment of a worker.

12. D: An approved respirator is an article of personal protective equipment; engineering, substitution, and administrative controls are not associated with PPE. Engineering controls consist of design modifications made to existing infrastructure and equipment to reduce worksite hazard exposure. Substitution controls merely work to replace hazardous materials and equipment with less hazardous options. Administrative controls use signage, training, and procedural changes to prevent or limit workers' exposure to hazards.

13. B: The subcontractor should rely on the knowledge and expertise of the contractor (contractually). The subcontractor should report any known or potential hazards to the contractor and host employer but should not implement controls. Finally, employers must protect employees from hazardous work conditions.

14. B: Keeping the load close to the body, maintaining a straight back, and using the two-handed grip are all part of sound lifting technique. The dominant hand is immaterial when lifting loads. The back is at its strongest when straight and all things—including bending the legs, maneuvering for proximity to the load, and body positioning—are done to ensure the back remains straight regardless of the body position.

15. D: Administrative controls place the focus of control on work behaviors, and redesign focuses on the source of the hazard. Choice *A* places control in the order of the hazard and work behaviors, Choice *B* places control in the order of the hazard and on the worker, and Choice *C* places both controls on or near the hazard.

16. B: A job safety analysis involves a written model for preplanning. Risk assessment matrices are used to determine level of hazard risk. A hierarchy of controls is used to eliminate or mitigate risk by prescribed effectiveness. An abilities assessment form is a checklist for obtaining yes/no answers to determine the performance level of systems and process where improvements can be made.

17. C: Risk matrices use event frequency vs. severity to determine risk level either quantitatively or qualitatively. Failure mode and effect analysis applies mathematical modeling to determine potential rates of failure and reliability for systems as materials and equipment. "Why?" methodology is used in incident investigation to determine root cause. Job safety analysis is used to detect hazards associated with job steps and apply corrective actions for each hazard.

18. C: Administrative controls include worker selection and training, isolation barriers to control worker movements and access, and signage to caution or warn workers to increase awareness and safe behaviors. Substitution controls are used as redesign and replacement of material, chemicals, and processes. Engineering includes the modification of materials and equipment to reduce hazard risk, while PPE is the equipment workers use to protect against hazards and is not a type of control.

19. B: The permissible exposure limit (PEL) is a standard and defining term utilized by OSHA to define worksite exposure limits to a substance in the air over a given span of time. Time-weighted average (TWA) is the exposure to any hazardous substance on the worksite over an eight-hour work day/forty-hour week, which if exceeded, mandates the worker exit the worksite. Short-term exposure limit (STEL) allows for fifteen-minute exposure increments (four times daily) if TWA is not exceeded; if exceeded, workers must exit the worksite for one hour. The rest are nonsensical terms.

20. A: A key element of inspections and preventive maintenance efforts is safety. Documentation of inspections, relative scheduling, safety controls applied, and tracking and monitoring for verification of performance level are reasons inspections should be led by the safety professional, although both upper management and qualified engineers might be involved in the inspections. The maintenance department is often unqualified to conduct safety inspections.

21. C: OSHA's Voluntary Protection Program comprises model organizations who exhibit excellence in safety performance. Organizations can reference OSHA VPP to enhance safety programs, promote safe work practices, and improve workplace safety culture. Preferential treatment during audits is not a benefit of these programs as it could compromise safety.

22. C: Concentration and duration describes the amount of the substance measured per unit combined with the time of exposure to the measured substance. Frequency and severity predict how often an effect is experienced with what level of effect. Cause and effect are relative to incident analysis. Physical and chemical are hazard forms.

23. C: Flammability and toxicity are associative chemical hazards. Autoignition temperature is a physical property. Flammability is a substance's likeliness to ignite or burn and cause combustion or fire. Toxicity is the term associated with human exposure to a toxin or poison (acute toxicity means a significant exposure over a short time period). Autoignition temperature is the minimum temperature at which a chemical or other substance can ignite without an external ignition. Irritability is an abnormal or excessive reaction in an organ or body part and can occur when a worker comes into contact with chemicals and substances, which are inhaled, absorbed, or swallowed. Coagulation is merely a result of blood desiccation.

24. B: Threshold limit values and recommended exposure limits are both recommendations and are NOT enforceable. An employer can, however, choose to protect workers with limits (TLVs and RELs) more stringent that OSHA PELs.

25. A: Workers are described as the backbone for safety success for their ability to determine if safety processes align with the work they perform. They are at the front line for detecting potential hazardous situations and reporting near misses. Safety professionals are the driving force of safety. Among the multitude of their responsibilities is effective communication of safety functions to all levels of an organization. The main role of management, along with the employer, is to show leadership and commitment in protecting employees from hazardous workplace conditions.

26. D: While increased awareness of potential hazards is a benefit of safety programs with cash incentives, over time, employees can begin to interpret cash bonuses as part of their regular income. In the event of an incident, employees will often conceal the event, if possible, to avoid what they perceive as a cut in pay for themselves as well as others. This results in ostensible lower incident rates, but the true result is loss of control. Alternative elements such as recognition awards (e.g., coats, hats) or gift cards can serve to curb nonreporting of incidents.

27. A: A greater number of workers and work disciplines increases the risks associated with simultaneous operations. No correlation exists between the size of an organization and the respective safety performance, safety culture, or quality of work.

28. B: The "Why?" method is used as part of incident investigations to determine the root cause. Incident investigations are reactive processes, meaning an event has already occurred (Why did this happen?). "What-if" analysis is a preplanning method that questions what could go wrong (What could happen?). Failure mode and effect analysis applies mathematical modeling to determine potential rates of failure and reliability for systems as materials and equipment. A safety culture analysis can be conducted via surveys, interviews, observations, data, etc., for the primary purpose of promoting shared responsibilities and safe work practices.

29. B: The time of day, worksite obstacles, or inclement weather may be factors, but consequence is the element combined with likelihood that creates injury risk. Likelihood means when and how often something will occur, whereas consequence means what and to what degree something will occur. Likelihood ranges from rare to almost certain, and consequence ranges from insignificant to catastrophic. The CHST must determine how to classify an injury risk and to find its mitigating measures.

30. D: Corrective actions must be tracked and monitored in order to determine the measure of effectiveness and to determine if additional or alternative actions are required. Analysis techniques such as root cause, "Why?" and "what-if" are used to determine what corrective actions may be appropriate.

31. C: At this stage of design (90 percent), a vast array of results and important information has been amassed. This stage provides a more complete look at the project as a whole and also allows for additional review of safety processes and design modifications that resulted from previous reviews and meetings.

32. C: Cost-benefit analysis considers the input of money and the level of effort put in for a corrective action and theorizes the point in time where a return on investment (or monetary balance) may be achieved. Design reviews, root cause, and "what-if" analyses do not consider such factors.

33. C: HSE signage, training, and safety notices are administrative measures, whereas equipment barriers illustrate engineering controls. OSHA will investigate catastrophes and fatal accidents when three or more workers are killed or hospitalized. The employer is obligated to report any such occurrence to OSHA within eight hours. Complaint/referral gives workers the right to lodge a request for worksite OSHA inspection when they believe they are in imminent danger, and this can be done confidentially.

34. D: Illnesses, injuries, and fatalities are compulsory reports. Arguments and personal confrontations are not.

35. C: The purpose of the chain of command is an efficient response from a special and streamlined group of individuals during an emergency. The level of title and years of experience (under normal conditions) are not factors for performance and efficiency in responding to emergency situations. Job responsibilities are also not a factor unless they would preclude the individual from being available to serve in emergency situations.

36. D: Employees will not take safety seriously without equal and consistent enforcement of safety and policy violations. Although effective safety committees, management knowledge, and positive safety cultures play critical roles in safety, without equal and consistent enforcement practices, safety programs, policies, and procedures are rendered meaningless.

37. B: As described below, for building design and construction, soil and topographical information is most relevant due to trenches and excavations, slopes and gradients, capacity to support footings, drainage ability and drainage systems, and water table and bedrock locations. The remaining categories can be useful but yield information more relevant to erosion, runoff, and local flood protections.

Soil Information:

- Determine the capacity to support parking lots, streets, and footings.
- Apply the slopes required in trenches and excavations.
- Reveal the level of the water table and location and depth of bedrock.
- Assess erodibility and drainage ability.

Vegetation:

- Identify isolated areas that are effective in resisting soil erosion.
- Consider vegetation ability to filter sediments and pollutants from storm water and runoff.
- Determine areas beneficial for aesthetics.

Topographical Information:

- Assess slope gradients critical for building designs and construction.
- Determine location and direction of drainage systems.
- Assess flow and capacity of surface water and storm runoff.

Hydrologic Data:

- Assess how water flows into and out of the site.
- Assess potential on-site flow volumes from upstream watersheds.
- Determine site drainage impacts regarding runoff and flooding of downstream water bodies.

38. C: OSHA 1926.1052(a)(2) states that stair slope or angle must fall between 30 and 50 degrees from horizontal. Steeper or higher angles result in tread overlap or require a diminished tread depth. Such elements reduce the safety for traversing stairs. Higher and lower angles represent components of ladders and ramps respectively.

39. D: Ladders must extend no less than 3 feet above an upper landing. In addition, extension ladders up to 36 feet require a minimum overlap of 3 feet, while those between 36 and 48 feet require at least 4 feet of overlap. Extension ladders between 48 and 60 feet (such as this ladder that extends a total of 50 feet) require a minimum overlap of 5 feet. Therefore, none of the answers are correct.

40. B: Only Choices *I* and *III* are associated with the goal of fall protection. Fall protection is considered an engineering control and PPE. As a reactant to a hazardous energy release (the fall and gravity), the goal is to protect workers by preventing them from hitting the ground.

Fall prevention equipment such as guardrails and restraints are associated with the goal of preventing the fall from occurring.

41. C: A diverse workforce can include personnel of different nationalities, which presents barriers to written and oral forms of communication. Signage uses universally accepted symbols and pictograms. An example of this widespread use is the Global Harmonization System of classification and labeling being implemented in many nations.

42. C: Many illnesses and injuries in the workplace/elsewhere can be treated and/or cured. Silica-associated illnesses and hearing damage (if allowed to occur) are simply irreversible once established.

43. D: Type A and stable rock have a higher compressive strength (> 1.5 tsf), which gives these soil types better cohesive properties. Higher cohesion results in a soil's ability to support itself under greater pressure. Type B soil's compressive strength does not exceed 1.5 tsf and is deemed unstable due to fissuring or having been previously disturbed. Less stable soils are less self-supportive and more likely to collapse with greater slope, lateral pressure, and trench depth. Type A soils can be sloped up to 53 degrees, and stable rock can be vertical.

44. A: This situation represents one of the elements of a permit-required confined space. With regard to entanglement and entrapment among all confined space types, these risks can never be completely eliminated. Each situation requires an analysis unique to the individual space. The introduction of equipment without analysis can exacerbate risk relative to inherent hazards such as space orientation, method of entry/exit, and the presence of piping and other fixed or moving parts. Acquiring a permit takes an array of hazard elements into account, where feasibility of equipment and various precautions and procedures are analyzed before they're prescribed. Without the elements provided by a permit, there is not enough information to support either Choice *B* or *C*.

45. C: This represents the soundest concept among the choices. All of the options suggest some amount of remaining risk. For Choice *A*, PPE should be viewed as a last resort, and cost alone should never rule the selection of controls. For Choice *B*, inherent risk can be mitigated, but it cannot be eliminated. For Choice *D*, with no controls in design or pre-planning, using a combination of (or multiple) controls suggests a high presence of armed hazards inherent within the process.

46. C: Falls, electrocution, and struck-by are described as primary physical hazards. A concussion is a condition resulting from exposure to a physical hazard.

47. B: Administrative, engineering, and substitution controls are elements of the control hierarchy, whereas operative is not.

48. C: Per 1926.100, OSHA adopts the criteria of the American National Standards Institute (ANSI). Additional elements (e.g., expiration dates and/or life of service, hat parts and insert replacements) lie in the specifications of the manufacturer. ASTM involves material and product properties and performance. NIOSH is primarily a research and recommending entity. UL is a safety and quality testing entity traditionally associated with electrical components.

49. B: Outdated work methods to maximize production were aligned with fitting the worker to the job. This philosophy maximized repetition and increased the risk to workers. The modern philosophy of ergonomics is to fit the job to the needs of the worker as evidenced in Choices *A*, *C*, and *D*.

50. A: The employer is responsible for selecting and prescribing PPE. The most effective controls for risk elimination are at the top of the hierarchy, since they address the hazard at the source. Furthermore, all higher options must be exhausted before considering PPE. For Choice *C*, PPE serves to protect the individual from the degree of consequence (severity), but it does not address hazard risk. For Choice *D*, PPE selection is based on several factors (e.g., worker size, appropriate material, protection level, allergy, increased stress, and reduced performance).

51. A: This situation suggests that HCP is not in place. These levels may not necessitate anything more than effective controls to reduce levels at the source. The information given does not suggest overexposure in an 8-hour workday (over 90 dBA), and NRR devices should not be applied to environmental levels (outside of the ear).

52. B: Ventilation, in concert with testing and monitoring, is an inexpensive and effective way to purge an environment of hazardous airborne particles. Time on-site controls including PEL, TWA, and STEL are to be utilized when the work locale cannot be safely purged of most suspended particles. Particulate respirators are the least expensive but do not protect against vapors, gases, or chemicals. Respirators work by filtering, purifying, or utilizing exterior air. Other respirators such as cartridge and gas mask respirators are used to filter or purify specific substances.

53. B: This applies to a side force applied within two inches of the top, and a down force applied to the top of the guardrail per OHSA 1926.502

54. C: An executive summary is condensed for the reader's convenience. Readers should have all the important information available, without the burden of poring through multiple pages and details.

55. A: A memo is used to introduce or lead actions that require subsequent attention. The content is not complex enough for report form and is too large for email.

56. C: As noted, hazard assessment and PPE are the responsibility of the employer. PPE should never be self-prescribed. In addition, this action is cause for concern because the employee is demonstrating that they're being exposed.

57. D: The presence of water is among the greatest hazards in excavation trenches. Threat of fissuring and collapse is magnified in less cohesive soils. Hazards are exacerbated with increased depth as lateral pressure increases where the soil can't support itself.

58. D: These are the characteristics that identify a confined space. A permit-required confined space must contain one of several possible hazards including hazardous atmospheres, engulfment, entanglement, and electrical hazards. Confined spaces present situations where workers are closely bound to potential hazards. Close proximity results in more easily triggered events that are difficult to isolate and evade. Confined spaces also magnify hazard risk level and duration of exposure.

59. A: FMEA involves prevention, pre-planning analysis, and other efforts relative to safety through design. The Domino and Chain of Events theories are "after the fact" measures that apply corrective actions to mitigate risk.

60. C: Severity, consequence, and likelihood of occurrence are all quantifiable items for assessed hazards. Impact on key performance indicators is more of a corporate concern.

61. A: Proper notification to affected personnel and machine shutdown are vital to initiating a lockout/tagout (LOTO) procedure. Effectiveness must be verified last via try-out.

62. D: Choices *A*, *B*, and *C* are administrative controls that modify human actions or movements. Job rotation modifies worker schedule and exposure time, monitoring inhibits movement or aids controlled access, and fences serve as isolation barriers to establish and maintain distance between people and material elements. Using a less hazardous chemical (in lieu of one more hazardous) is an example of substitution and is a higher control among the choices.

63. C: Excavations greater than 20 feet require a professional registered engineer. OSHA 1926.652, 1926 Subpart P App B, and other parts express the requirements for registered engineers with respect to protective systems. Complexities in required design are proportional to trench depth. This means that as trench depth increases, the compounding dynamics of the soil's weight compounds the lateral pressure forces that test a soil's compressive strength, in contrast to trenches of lesser depth. Increased depth also increases the potential for the presence of water, which results in soil fissuring, compromising sidewall integrity. Trench failure at such depths requires ensured effectiveness in design that exceeds the capabilities of a non-engineer.

64. D: Management commitment must be in place. Management develops and maintains programs, grants approvals for corrective actions and associated costs, and has the most control over elements contained within Choices *A, B,* and *C,* among many others.

65. D: Other processes are performed for many reasons. The performance of a JSA, Safety Audit, and CBA are not indicative of a reaction to an incident. Audits test performance and compliance to programs and standards. A CBA is primarily a spending based initiative, and a JSA is performed initially to prescribe procedures and detect associated job hazards. Determination of root cause, prevention of recurrence, and fulfilling legal requirements all point to a process directly reactive to an incident. The results of an incident investigation could, however, spur the actuation of a JSA revision, exterior (OSHA) audit, or CBA as the extension of an improvement/recommendation expenditure.

66. D: Oxygen levels are approximately 21 percent in normal air/atmosphere. Lower oxygen levels result in negative effects on human health. Oxygen levels outside the normal range also effect LEL and UEL with respect to combustion and explosion.

67. D: Rope inspections are compulsory to ensure that there is no damage, while crane hook and chain and load ratings must be visible. Air conditioning is not mandatory, although it is a very good idea in hot weather.

68. C: This communication method increases process safety through awareness. Three-way communication is exclusive to verbal command and the responses of those involved in or near-to the activity. Hand and arm signals are used where interference and deficiencies exist in line-of-sight or audible detection. Pictograms are visual modes to increase comprehension.

69. A: A tie-in is required at a 4:1 ratio or greater. A unit that is 30 feet tall and 6 feet wide has a height to width ratio greater than 4:1. Working areas must be fully planked; however, planking must be at least 18 inches wide for walking/traversing. In addition, asphalt will crack and sink under heavy pressure, which can cause the scaffold to collapse.

70. B: The combination of a hazard and exposure to them results in accidents. Barriers and PPE are hazard controls. Chemicals may be a factor in an accident, but exposure to a hazard is what causes accidents. The number of workers present may affect the exposure to a hazard, but does not directly cause accidents. Working from heights is a component of fall hazards. Restricting areas is an aspect of administrative control.

71. C: Mitigation, categorization, and assignment are post-hazard assessment activities. A hazard is first identified, and then assessed; it is then categorized in respect to likelihood and consequence and assigned to the individual or group responsible for its timely mitigation.

72. B: Per 1926.651 for safe means of egress (reduced distance of lateral travel in any direction), 25 feet of lateral travel is the maximum distance permissible by ladders. This is true for stairs, ramps, or any other means used in trenches greater than 4-feet deep.

73. A: Although all choices represent hazards, air quality has been (and continues to be) the greatest hazard for confined space workers.

74. D: Among the choices, only guardrails can prevent the fall from occurring. All of the other options address the hazard after the fall or have elements that can be defeated. Fall arrest systems are designed to prevent the worker from hitting the ground but can still result in injury.

75. A: Ten percent denotes personnel not on-site or directly involved with an incident, so it will be a low percentage figure. It is generally accepted that 90 percent of worksite incidents are caused by direct worksite behavioral actions on the workers' part. These are actions directly and locally affiliated with the incident. An example would be a worker operating the crane who was visibly upset when arriving for the shift, and it is thought this may have caused the operator to over-swing the load and impact a truck. Indirect involvement might have been the worker who mistakenly packed the wrong tensile strength wire rope for the crane back at the warehouse and has never visited the worksite but still affected the incident's outcome.

76. D: All choices represent a sound concept for hierarchal control applications. Higher controls address materials and equipment, while lower controls address the person and human actions.

77. C: A Job Safety Analysis (JSA) lists job steps and associated hazards and suggests corrective actions and procedures. A JSA can be developed before the start of a new task or be reviewed and revised as a reaction to an incident.

78. C: Information gathered via this process can be used to determine the strengths and weaknesses of the culture, policies, and programs. Compilation of scores and the deviation of averages indicate the elements that can be addressed by process or department.

79. B: The attendant is NOT responsible for retrieval systems and feasibility. This responsibility belongs to the supervisor. With training, the supervisor can also act as the attendant or entrant. The attendant has many responsibilities that are vital to safety for confined space activities. The attendant must be fully dedicated to these duties and cannot perform any other duty unless relieved by another qualified attendant.

80. C: OSHA's rules for housekeeping and all other provisions that apply to all construction employers can be found in 1926, subpart C. While a sloppy workplace can definitely lead to fall scenarios, 1926, subpart M, applies specifically to falls from heights. OSHA's 1910 standards address safety and health guidelines for the "general industry" employers such as manufacturing plants, hotels, and all other environments not involving construction activity; 1926 is where safety standards for construction activities can be found.

81. D: The point of safety training is to develop new skills that will result in improved safe work performance. While it is important to retain training documentation for human resources and personnel considerations, the documentation itself does not guarantee safe work practices. OSHA generally does not require training documentation; OSHA bases its evaluation of whether employees have been adequately trained to perform their jobs safely by observing their work practices. A course outline or description from the instructor should be retained with training documentation, but a very descriptive and correct one does not necessarily imply that a particular course is the best option. A training-needs analysis will help to determine what training a CHST should recommend. Employer-provided safety training is not always the best option. Only the employer can provide guidance on its safety policies and procedures, but sometimes third-party trainers are more appropriate for specialty hazard training, such as chemical safety or new lead remediation techniques.

82. A: Operator manuals are an integral part of the documentation for operational purposes and must be included in the paperwork for cranes during an audit. There are no paint codes for hook and chain. Load limits and crane rope/wire will undergo load testing during an audit but do not fall into the mandatory documentation for an audit.

83. A: The Board of Certified Safety Professionals believes that hazard control is most effective when the immediate supervisor is responsible for correcting identified hazards. The OSHA 301 is an incident report form used for recordkeeping purposes following injury events. Weekly inspections of the job site are far too infrequent to ensure adequate hazard control. Whether documented or not, hazard identification must take place continuously so that hazards are promptly corrected. This can be as simple as an employee picking up a hammer that was left along a common travel path and noting it down for the supervisor to mention at the end- of-the-day's safety meeting. Work stoppages are well advised for high-gravity hazards, those that repeat themselves, and in many other situations, but stopping all activity on the site is not always the best course of action. For example, the hammer that was dropped and could have caused a tripping incident may be worth mentioning at the end-of-the-day's meeting but would not be such a critical situation that all work activity should be stopped to address it.

84. C: Task-specific hazards, SDS, and worker safety feedback are all operational safety ingredients pertaining to a JHA, but workers' compensation is associated with the human resources department.

85. C: OSHA has an inherent right to inspections, document review, and the institution of voluntary worksite safety programs, but their authority does not extend beyond the workplace.

86. B: 70 to 100 milliamps are generally recognized as the amount of current that should be considered fatal. 7 milliamps are not a fatal amount of current, but a contact will be noticeable and is the upper limit of when a GFCI system will trip when functioning properly. 7 and 70 amps are far greater than the amount of current that can be expected to cause a fatality.

87. C: Running cords through doorways is permissible so long as cords are protected from damage. Running cords through holes in walls is never acceptable. This is commonly done to reduce the length of cord run across the ground, which will reduce a tripping hazard but create the even greater hazard of electrocution. Insulation can be torn off from unseen sharp objects inside the wall or on the other side. A CHST should never create a greater hazard when eliminating others. The use of staples, nails, and any objects that could damage electrical cords should not be used as a means of routing, nor should they be allowed to contact cords. Electrical tape should not be used to repair electrical cords; a properly repaired cord will look like any new, undamaged cord and repaired with similar "factory" components.

88. D: The only exception to the requirement to provide a continuous path to ground in construction is for tools that are double insulated. Double-insulated tools are indicated by the symbol of a square inside another square.

89. D: Crocidolite, chrysotile, and amosite are forms of asbestos, and vermiculite is not. Used in coatings, cements, and insulators for high-heat environments such as steam engines, Crocidolite is blue and has the highest heat resistance. Chrysotile is found in floors, walls, roofs, and ceilings of residential and commercial edifices and is the most common form of asbestos. It is also found in brake linings, gaskets, seals, and insulation. Amosite is amphibole-type asbestos, as it has needle-like fibers. Known as brown asbestos, it originates mainly in Africa. Used in pipe insulation and cement, it is also found in ceiling tiles and insulation boards with high asbestos content (up to 40 percent). Vermiculite is a hydrous silicate material that expands when heated, although it is not a form of asbestos.

90. A: GFCI will interrupt current when a disparity of 5 to 7 mA is detected between the hot and neutral return wires and will take place within 0.025 second (1/40 of a second) when functioning properly. An assured equipment grounding conductor program is an exhaustive system of documentation and testing that is the only alternative to using GFCI in construction. A power cord does not need a ground prong for GFCI to function correctly, since the ground wire plays no part in the operation of the GFCI, so even non-grounded power tools that are double insulated can be protected by GFCI.

91. C: Fall protection, typically a personal fall arrest system (PFAS), is always required when working on aerial lifts; the lift platform does not have to be 6 or more feet from the ground. Although not treated as scaffolds, OSHA's aerial lift requirements can also be found in 1926, subpart L, Scaffolds. OSHA does not require that fall protection be provided on scaffolds until 10 feet above the ground or the next lower working surface, though it should be provided whenever feasible to do so. OSHA's general duty clause has nothing to do with specific stairway and ladder safety requirements; the general duty clause is a stipulation that an employer can be cited for an employee hazard exposure even when there is no specific standard that addresses the particular hazard in question.

92. A: Standard guardrail systems are considered a fall restraint because they prevent a fall from happening. A safety net is an arrest system. A shock-absorbing lanyard and full-body harness are both components of a personal fall arrest system (PFAS). Restraint systems are generally preferred to arrest systems for several reasons—most importantly because a fall is not possible—and are recommended whenever practicable.

93. B: The maximum total fall distance allowed in a personal fall arrest system is 6 feet; this includes all attached components and is not based on the length of the lanyard alone. A PFAS must utilize a full-body harness; body belts are never allowed with fall arrest systems. Fall restraint is always preferred to fall arrest, since no fall can occur. However, sometimes restraint systems are not possible, and the contractor has no choice and must employ a PFAS. PFAS components must meet OSHA's strength requirements outlined in 1926, subpart M, but OSHA itself does not perform product safety testing or approve any person, place, or thing.

94. B: Regardless of the method chosen, employees must always be notified of and expected to eliminate the hazard promptly and correctly; this is the case for floor holes and for any other hazard on a construction site. A guardrail system is an acceptable option but is not mandatory. If the hole is too small for a person to fall through, then a guardrail system would not be advised; it is overly complicated and time-consuming. Flagging or caution lines are not acceptable. Covers must be strong enough to support at least twice the maximum axle load of the largest vehicle expected to travel over it; there is no fixed strength requirement.

95. D: OSHA requires that the top rail of a standard scaffold guardrail system be installed between 38 and 45 inches from the deck. For other horizontal working surfaces in construction, such as a rooftop, the top rail must be installed between 39 and 45 inches. Fall protection on scaffolds is not mandatory until when more than 10 feet above the ground or the next lower working surface exists. OSHA's general rule to provide fall protection at 6 feet in construction does not apply to scaffolds. OSHA allows a scaffold's cross brace to take the place of one, but not both, of the guardrails. Employees working below others should not forego the use of hard hats when toe boards are used on scaffolds because toe boards will not prevent all objects from falling; hard hats are mandatory whenever there is a falling-object hazard.

96. C: The results of a BBS in the worksite—if successful—will cause a decrease, not increase, in worksite incidents. Direct contact between the workers and OSHA is not necessarily associated with a successful BBS.

97. D: An antecedent (meaning before) is an action taken or preexisting condition that leads to an accident later. Humans are sentient and emotional creatures, making it necessary to be observant of at-risk behaviors (behaviors causing actions that increase the risk of injury or illness). Consequence is the result of a behavioral action by a worker or workers. Behavior means what people do and why they do it, and a Behavioral-Based Safety Program is used to improve what people do by behavioral improvement.

98. C: OSHA permits the use of a personal fall restraint system (PFRS) utilizing a body belt when a fall from an aerial lift is not possible, though this practice is generally discouraged in favor of a personal fall arrest system (PFAS). PFAS components are stronger and more widely available than those used for restraint and should be seen as providing a higher degree of safety and reliability. Another consideration when introducing body belts, even when properly used in restraint situations, is that employees may see them used and incorrectly assume that they are acceptable in fall arrest situations and utilize them as such. OSHA does not permit the use of guardrails as a form of fall protection for aerial lifts. A suggestion to require that employees working over water in an aerial lift wear a fall or restraint harness would not be well advised because the drowning hazard may be greater than the fall hazard. When correcting hazards in construction, a CHST must avoid the introduction of even greater hazards in the process and should always consider all options available to minimize risk. OSHA allows up to a 6-foot drop in a PFAS, and this should always be minimized, since an arrested fall can still cause injuries to the back or other body parts. When calculating the maximum arrest distance, the total length of all PFAS components must be addressed, not just the lanyard.

99. A: OSHA evaluates the effectiveness of safety training based on an observation of work practices; while important for many reasons, training documentation itself ultimately plays no role in the evaluation. A CHST should consider taking the same approach. If training does not result in improved safe work performance, then training was not a need, and other factors leading to a lack of safe work performance will need to be identified. Defined by OSHA, a competent person is "... one who is capable of identifying existing and predictable hazards in the surroundings or working conditions which are unsanitary, hazardous, or dangerous to employees, and who has authorization to take prompt corrective measures to eliminate them." OSHA's definition applies mostly to supervisors, but any sufficiently capable employee can be deemed a competent person, and there can be any number of them at a construction site. Electrical cord repair does not need to be undertaken by a licensed electrician, but the cord must be repaired such that it is equivalent to an undamaged cord that one can buy new. A personal fall arrest systems (PFAS) is not mandatory on scissor lifts; as with other types of scaffolds, guardrail systems are acceptable and very common.

100. C: Imminent danger, catastrophes, and fatal accidents and complaint/referral constitute OSHA methods of inspection, although supplemental inspections are not included in this protocol. Imminent danger means the employer will be asked to immediately remove workers from the worksite, as there are hazards that require instant corrective action and cannot be mitigated through normal channels.

101. B: Choice *A* is undesirable because a minor safety hazard should not trigger a site-wide work stoppage. Although it is advised to err on the safe side when in doubt, mandatory work stoppages are most appropriate when a high-gravity scenario is observed, such as an employee almost being crushed by a front-end loader or an aerial lift being operated in a dangerous manner; these scenarios require immediate attention. Choice *B* is the best option because hazard control is most effective when the immediate supervisor is responsible for correcting identified hazards. In this case, the hazard has repeated itself; the CHST should submit a report to senior management promptly so that appropriate action can be taken. Choice *C* is not appropriate because a CHST should not be doling out disciplinary action; the role of a CHST is to assist construction managers to make decisions leading to a safe work environment and minimizing potential loss events, not disciplining employees. The actions taken in Choice *C* will ensure that more appropriate parties than the hazard-control advisor will determine which actions to take, which may be disciplinary in nature. Choice *D* is not advised for reasons previously addressed.

102. C: OSHA requires that first-aid kits in construction be inspected at least once per week and also prior to assignment to a site.

103. D: "Universal Precautions" refers to the practice of treating all bodily fluids as if they contain blood-borne pathogens, which is planning for the worst. This is a highly recommended practice, since it is impossible to know whether blood or other potentially infectious materials contain pathogens without access to employee medical data, which a first-aid responder will typically not have, and also because an employee may have a blood-borne disease and be unaware of it.

104. C: Microwave ovens, alpha particles, and beta particles denote non-ionized radioactivity, whereas gamma rays are ionized radiation, which is potentially very harmful to humans.

105. C: In an emergency, the first priority is to safeguard personnel, equipment, and processes, in that order. Relevant emergency management personnel should be advised of the situation as soon as possible, but ensuring that personnel are safe is always the most important consideration during an emergency. OSHA's requirements for emergency actions plans (EAPs) are outlined in 1926, subpart C, General Safety and Health Provisions, and also 1910, subpart E, Means of Egress. OSHA's Hazardous Waste Operations and Emergency Response requirements are found in 1910, subpart H, Hazardous Materials. There is no requirement that alarms must be 105 A-weighted decibels, but any alarms, when used, should be loud enough that they will be heard by all employees who will need to hear them.

106. A: Class B fires involve Flammable Liquids and Gases. Ordinary Combustible fires are Class A, Electrical Equipment fires are Class C, and Class D fires involve Combustible Metals.

107. B: On a modern construction site, *dry chemical* and *multipurpose dry chemical* extinguishers are the same, and both phrases are used. Prior to the development of modern "multipurpose" dry chemical extinguishing agents, which are suitable for Class A, B, and C fires, the first dry chemical extinguishers were suitable only for Class B and C fires and have become unpopular. Fire extinguishers on construction sites should be inspected at least monthly; this is an NFPA (National Fire Protection Association) consensus standard that OSHA has adopted and enforces. An extinguisher must be located within 50 feet of 5 or more gallons of flammable materials in an outdoor construction environment, not 10 feet. There is no mandatory fire extinguisher to employee ratio, such as one extinguisher for every ten employees.

108. C: Training topics must include identification and knowledge of blood-borne pathogen hazards, control methods, emergency actions, personal protective equipment (PPE), the hepatitis B vaccine, the employer's exposure control plan, and general housekeeping considerations. Basic first aid, cardiopulmonary resuscitation (CPR), and automated external defibrillator (AED) are not topics addressed by the blood-borne pathogen standard, though personnel tasked with duties involving these topics would typically be required to receive blood-borne pathogens training. First-aid training is acceptable to OSHA if equivalent to Red Cross or Bureau of Mines curricula and is readily available in many locations; no such provision applies to blood-borne pathogens training. OSHA's blood-borne pathogens standard is found in 1910, subpart Z, Toxic and Hazardous Substances.

109. D: During construction activity in multistory buildings, there must be at least one fire extinguisher on each floor, and of those, at least one must be adjacent to a stairway. Evacuation assembly points do not need to be fixed for the duration of a construction project; assembly locations may need to change during the course of work, and multiple assembly locations may be preferred when employees will be working at multiple places on the site during the workday. Employers with ten or fewer employees are not exempt from OSHA's requirement to develop an emergency action plan (EAP). OSHA's blood-borne pathogens standard requires employers to make the hepatitis B vaccine available to all employees with job duties involving a reasonable exposure to blood and other potentially infectious materials, such as first-aid responders. Although it is not mandatory that employees receive the vaccine, employers must make it available to them and pay for it when they elect to receive it.

110. D: The objective of an incident investigation is to prevent a similar incident from occurring. Arriving at an incident site in a safe manner and caring for the wounded must take place before the fact-finding begins, but these actions are not objectives of the investigation itself. Liability, regulatory, and legal matters will always be topics of concern during an incident investigation process, but they are of secondary importance to the primary goal of ensuring the incident does not repeat itself.

111. B: A safety training matrix tracks mitigation, performance, and personnel assigned to the resolution of safety issues. Standard operational procedures are to be monitored by CHST to ensure they are being executed correctly.

112. B: The best place to interview a witness is the incident site, but it will not be an appropriate interview venue when the incident is very bloody or particularly ghastly in any regard because many employees will be upset by this and will not want to be on the site. Witnesses must understand that the fact-finding interview is NOT a disciplinary interview, and so this should be communicated to them prior to asking any questions; a proper investigation identifies the root causes of the incident through fact-finding, not fault-finding. Interviews should take place as soon as possible following an incident because important details may be forgotten, and the general clarity of information recall will degrade quickly in the first few days following an incident. An administrator's office is not an advised interview location because it may put the witness on guard.

113. A: When beginning an incident investigation, the priorities are to arrive safely, size up the situation, care for the injured, and lastly, protect the property. A safety administrator should not be the lead person in routine incident investigations; the most immediate supervisor, such as a supervisor, is the lead person in most routine incident investigations. There are frequently multiple causal factors in an incident. Actions such as corrective measures, disciplinary proceedings, and meetings with union representatives should only take place once useful factual data has been gathered.

114. C: The number of employees handling evidence should be kept to a minimum to reduce the likelihood of errors and mishaps in the transfer process. Anonymity in the evidentiary chain-of-custody process is the opposite of what is desired; without knowing who handled the evidence, it is impossible to know when or where the evidence may have been mishandled and perhaps lost. All employees assuming custody should be required to sign a chain-of-custody document and should receive a receipt of the transfer. General contractors and subcontractors should always work together to develop procedures common to operations that will involve both parties and should coordinate to the greatest extent possible in order to ensure a productive and safe work environment. When others already know the information that is being presented, it is a safe bet that they will notify the presenter of this to avoid wasting time. But the opposite situation is impossible—they cannot know what they do not know, and so a CHST should avoid being overly presumptuous when presenting information so that important details are not left out.

115. C: The OSHA Form 301 is the employer's Injury and Illness Incident Report that must be completed within 7 calendar days of the day of the incident. The OSHA Form 300 is the Log of Work-Related Injuries and Illnesses, and the OSHA Form 300A is the Summary of Work-Related Injuries and Illnesses.

116. A: With the exception of the tetanus vaccination, rabies and all other inoculations following an incident will trigger OSHA recordkeeping duties; the tetanus vaccination is included in the 1904 standard's list of basic first aid and does not trigger OSHA recordkeeping. Blood tests are considered diagnostic procedures and alone will not trigger OSHA recordkeeping. However, if a blood test reveals an employee has incurred a blood-borne pathogen, for example, hepatitis C, then the incident must be recorded because hepatitis C is a significant illness that requires entry on the OSHA 300 log.

117. B: Choice *B* is the sole answer associated with the safety and impartiality a CHST must exercise in the worksite.

118. D: General contractors do not record the injuries of subcontractor employees. Incidents involving a work-related fatality, days away from work, modified duty such as work restrictions and temporary job transfers, and cases involving medical treatment not identified as basic first aid are all recordable events. OSHA's 1904 standard includes the practice of draining fluid from a blister as basic first aid.

119. B: The TRC rate was 5.00 in both 2014 and 2015 and increased to 5.49 in 2016. This formula is used to calculate incident rates: (*N* x 200,000)/EH. *N* represents the number of cases relevant to the particular incident rate desired. *EH* stands for "Employee Hours," which is the total hours worked by all employees in the reporting year. EH will be the same regardless of which incident rate is calculated. 200,000 is a fixed constant that never changes.

120. B: The DART rate was 2.75 in 2016. This formula is used to calculate incident rates: (*N* x 200,000)/EH.

121. A: While OSHA allows an employer up to 7 days following the day of the incident to complete the 301-incident report, it should be completed as soon as possible, since witnesses may forget important details. Recording incidents on the OSHA recordkeeping forms that do not meet the criteria for OSHA recordkeeping should be avoided; this practice will artificially inflate incident rates and could disqualify a contractor from a bid. Logging as many incidents as possible, including near misses, will maximize the efficacy of trend analysis, but non-recordable events should never appear on the OSHA forms. The OSHA Form 301 is not limited to medical treatment events but is used for all events meeting the criteria for OSHA recordkeeping entry. Much of the data that is eventually entered on the OSHA recordkeeping forms is generated by field staff, especially the initial incident data for use on the 301; these duties should not be limited to safety personnel, although they will be responsible for ensuring that OSHA recordkeeping data is correct and that forms have been completed properly.

122. C: A scaffold must be secured from accidental displacement when its height is more than four times greater than its width at the base. Common methods of achieving this are to use outriggers or guy wires or to fasten the scaffold to an adjacent surface, such as a brick wall, on the interior side of the scaffold. An administrator's office is not an advised interview location; it is not a neutral venue and may put the witness on guard. OSHA does not mandate that inspections be documented, only that identified hazards be corrected. Class D fires involve Combustible Metals and require special extinguishers that are only appropriate for Class D fires.

123. D: OSHA's recordkeeping standard is found at 29 CFR 1904. Maritime, general industry, and construction standards can be found at 29 CFR 1915, 1910, and 1926, respectively.

124. C: Rolling objects, swinging objects, and falling objects constitute a struck-by scenario; being struck by a coworker constitutes a worksite altercation.

125. C: When working with hot tools, sparks can fly up to 35 feet and ignite flammable gas in the air or get caught on clothing or in cracks in the walls and floors.

126. B: Back injury and other types of sprains or strains are considered physical injuries. Struck-by (Choice *A*) is the hazard of objects striking workers. Health hazards (Choice *C*) are those in which toxic substances or other chemicals can cause illness. Stress (Choice *D)* is not a specific type of hazard, though stress injuries from repetitive movement are considered physical hazards.

127. B: Powder-actuated tools use firearm cartridges with explosive charges to propel fasteners into hard substrates like concrete and steel, so the correct answer is Choice *B*. However, they cannot go off and shoot nails (Choice *A)* or be fired at a distance (Choice *D)*, because they have a safety mechanism that requires the muzzle to be pressed against a hard surface before it will fire. High-velocity powder-actuated tools that fire over approx. 328 ft./s are not sold in the United States.

128. D: Neither heat rash (Choice *A*) nor heat stroke (Choice *B)* cause permanent tissue damage. Heat rash is caused by clogged sweat glands and can heal without permanent damage. Heat stroke is characterized by headache, vomiting, dizziness, and, eventually, passing out.

129. B: Electric shock is considered a physical hazard. Health hazards (Choice *A*) are those that cause illness, and environmental hazards (Choice *C)* are those that cause damage to the environment. Imminent danger (Choice *D)* is not a hazard classified by GHS, and any hazard could become an imminent danger if it is not prevented.

130. D: The span of control limits the roles and responsibilities of any one agency. Modular organization (Choice *A)* refers to mobilizing agencies only if their role is needed in a particular emergency. Unit of control (Choice *C)* refers to the central leadership of the incident command system. Management by module (Choice *B)* is not a key concept of incident command system.

131. C: Builder's risk covers damage to the property, equipment, or materials due to natural disasters or during normal work, but it does not cover floods (Choice *A*) or faulty workmanship (Choice *D)*. General liability covers accidents on the job (Choice *B*) and is a separate policy from builder's risk.

132. A: Suppliers lists, customer lists, blueprints, and product specs are examples of trade secrets. Procurement documents (Choice *D)* are considered only confidential information. Employee addresses (Choice *B)* are personally identifiable information, and medical records (Choice *C*) are confidential data protected by HIPAA.

133. C: BCSP requires certified professionals to conduct relationships with integrity to avoid conflicts of interest and report misconduct from any certified professional. Choices *A*, *B*, and *D* are all included in the BCSP code of ethics but do not relate directly to conflicts of interest.

134. D: During an individual assessment, leaders assess groups and individuals to determine which skills they have and which skills they need to perform at the desired level determined by the organizational assessment (Choice *C*). The training needs analysis (Choice *A*) determines how to deliver the training to the workforce based on their limitations. The workforce assessment (Choice *B*) refers to the individual assessment.

Dear CHST Test Taker,

We would like to start by thanking you for purchasing this study guide for your CHST exam. We hope that we exceeded your expectations.

Our goal in creating this study guide was to cover all of the topics that you will see on the test. We also strove to make our practice questions as similar as possible to what you will encounter on test day. With that being said, if you found something that you feel was not up to your standards, please send us an email and let us know.

We would also like to let you know about another book in our catalog that may interest you.

Six Sigma Green Belt

This can be found on Amazon: amazon.com/dp/1628454164

We have study guides in a wide variety of fields. If the one you are looking for isn't listed above, then try searching for it on Amazon or send us an email.

Thanks Again and Happy Testing!
Product Development Team
info@studyguideteam.com

www.ingramcontent.com/pod-product-compliance
Lightning Source LLC
LaVergne TN
LVHW082344200926
848481LV00012B/3441
* 9 7 8 1 6 2 8 4 5 7 0 1 8 *